Cerebrum 2010

Cerebrum 2010

EMERGING IDEAS IN BRAIN SCIENCE

Dan Gordon, Editor

DANA
PRESS

New York • Washington, DC

The Dana Foundation
745 Fifth Avenue, Suite 900
New York, NY 10151

DANA is a federally registered trademark.

Printed in the United States of America

10 9 8 7 6 5 4 3 2 1

ISBN-13: 978-1-932594-49-2
ISSN: 1524-6205

Art direction by Kenneth Krattenmaker.
Cover design and layout by William Stilwell.
Cover illustration by istockphoto / ktsimage.

William Safire (1929–2009)

Cerebrum 2010 is dedicated to the memory of William Safire, chairman of The Dana Foundation, who passed away as this volume was in final editing. In 1998, as a board member fascinated by the neuroscience Dana was supporting, Safire conceived the quarterly print journal, *Cerebrum*, and persuaded the late David Mahoney, then chairman, to launch it. Its mission: to provide a forum on the changes brain research promised not only in human health but also in important aspects of everyday life. As Dana chairman, Safire saw the immediacy and dynamism of Internet communication as tailor-made for *Cerebrum* and, in 2006, had the journal moved online. For 12 years, until just weeks before his death, he guided every article, including those in this edition, demanding clarity, respect for plain language, and above all a point of view that readers could engage with and argue about. Whether the topic was the science itself or its multiplying implications for society in areas as different as ethics, the arts, law and education, he wanted people to care about it.

—The Dana Foundation

Contents

Book Reviews

Foreword

By Benjamin S. Carson Sr., M.D.

Benjamin S. Carson Sr., M.D., is the director of pediatric neurosurgery at Johns Hopkins Children's Center and a professor of neurological surgery, oncology, plastic surgery and pediatrics at the Johns Hopkins Medical Institutions. He holds the inaugural Benjamin S. Carson Sr., M.D., and Dr. Evelyn Spiro, R.N., Professorship in Pediatric Neurosurgery. Carson has written four books, including his autobiography, *Gifted Hands: The Ben Carson Story*. He was awarded the Presidential Medal of Freedom in 2008.

CALL ME BIASED, but I think that neuroscience is by far the most interesting and stimulating area of science. When I was in medical school, cardiovascular surgery, a field in which doctors were making lifesaving break-throughs, looked for a time like it would be my specialty. But then I started delving deeply into neuro- and behavioral science and became intrigued with the things that neurosurgeons were able to do. I quickly learned that even though we had vast knowledge of the brain, neurology was an area with even greater potential for discovery and hope. I abandoned my fleeting fascination with cardiovascular surgery and directed my attention to neuro-surgery, where every day that promise of discovery seems to come true.

In 1985, after I performed my first cerebral hemispherectomy—removal of one of the brain's hemispheres—on a child who had intractable epilepsy, there was a flurry of media attention. It was not the first such procedure, but it is a rather radical operation that largely had fallen out of favor because of its associated complications. Hemispherectomy had become so poorly regarded that, when the news reports came out, three medical students at a prominent school in another part of the country called to ask if the stories were true. They said they had tried to verify the reports with their own institution's neurosurgeons, who told them that they must have misunderstood.

That was a quarter of a century—and more than a hundred hemi-spherectomies—ago for me. Since then, thanks to techniques refined to decrease the surgery's complications and mortality rates, neurosurgeons have performed thousands of hemispherectomies at children's centers across the country. More important, we have a much better understanding of why such a radical operation can be successful. It is because the imma-ture brain is at a stage of development when its plasticity—the ability to change and adapt—is at its most vigorous. And this is just one example of the promise of neuroscience: As we continue to explore the concept of plasticity, we will perhaps discover how to create it in mature brains. This, in turn, will have a profound effect upon brains that have been trauma-tized through head injury, vascular accidents or a myriad of other causes.

The specialty of neurosurgery has changed dramatically during the decades I have been involved with it. We have been able to take great advantage of advances in both brain imaging and neurophysiological electrical monitoring to perform delicate operations that nobody had dreamed of when I entered the field. Some people now predict that neurosurgeons will become obsolete because doctors will be able to attack all health problems at the molecular level. They imagine a scenario like one portrayed in the movie *Star Trek III*, in which the starship crew discovers a human skull from way back in the 1990s that shows evidence of a craniotomy and are astonished that people were once so barbaric as to actually cut into the human skull and the brain.

I think it is premature to predict the demise of this aspect of neuroscience, but the fictional crew's reaction is based on a kernel of truth: Already, thanks to technological advances, we are able to do with minimally invasive techniques the same kinds of procedures that were massive undertakings not many years ago. Yet, as we discover more about the brain, more reasons for potential surgical intervention emerge. For example, researchers are studying very specialized sensors that can detect, monitor and analyze impulses from a single neuron and integrate those impulses with others to communicate with electronic devices. One day such sensory systems may enable patients with quadriplegia to drive a wheelchair, or potentially even a car, by using their brain impulses. It will not be a jump of unimaginable magnitude to conclude, then, that patients could use those same brain impulses to stimulate neuromuscular tissue and use their extremities once again. To those freshly minted doctors in neurosurgery residencies, do not fret: Those sensors will have to be implanted in the brain, so you will have work to do.

I do believe that many of the things neurosurgeons do now will no longer be necessary as we gain even more knowledge about the molecular basis of disease. Recently, when I was in Oklahoma to give a public talk, a family approached me and asked if I recognized the 20-year-old man who was with them. He was a handsome youth, quite articulate and polite. The family informed me that I had operated on the man when he was 6 months old and had a very large tumor of the brain stem. Doctors had given him six months to live at the time he came to me. We performed a rather extensive operation but were unable to remove the tumor completely.

Surprisingly, the remaining portion of the tumor spontaneously regressed during the next couple of years, and eventually there was no radiological evidence of it. Other cases of spontaneous regression of tumors have been reported, and it appears that the immune system plays a significant role in that regression. The point here is that advances on the non-surgical front will eventually eliminate the need for surgical cures for many problems in the central nervous system and other parts of the body.

In addition to my clinical responsibilities, I have the privilege of giving many public lectures, and those have taught me one more delightful thing about neuroscience: Almost everyone, from kindergarteners to the most senior of citizens, is fascinated by the human brain and its potential. Even though knowledge about every aspect of the brain has exploded in the last couple of decades, each new discovery exposes another cadre of questions to be answered. In a way this makes neuroscience even more exciting because we know we will be in a permanent mode of discovery.

This edition of *Cerebrum* embodies that truth; it is full of surprising discoveries and provocative questions about how and why our brains work the way they do. For example, we look at advances in neuroimaging and what it can and cannot tell us, a topic that is particularly interesting to a neurosurgeon. I have a program at the hospital in which, each month, I talk to about 800 schoolchildren about the brain and human potential. Frequently, I tell them about the wonders of functional magnetic resonance imaging and positron emission tomography scanning and how these technologies allow us to correlate brain activity and physical functions. We can now determine which parts of the brain are at work not only during certain activities, but also during expression of certain emotions. One day we may even be able to link certain brain activity with particular types of thoughts. But the article "Neuroimaging: Separating the Promise from the Pipe Dreams," by Russell A. Poldrack, splashes a bit of cold water on one of my favorite moments when talking to these young students—when I tell them it is probably a good idea to start thinking pure thoughts now, because in the future people might be able to detect lies and other thought patterns without our consent. Poldrack says that's a stretch.

Another subject that has inspired a great deal of talk in recent years is how to nurture the developing brain. In this volume we will see several sides of this fascinating topic, including the nutritional aspects of brain development and how continuing brain development affects learning in adolescence and why teens are such risk-takers. The chapter on video games should be alluring if you play these games or have children who do, since we have long suspected that along with some benefits, there might be some deleterious effects. I am struck by the number of parents who ask me if it is OK to put their kids on various behavior-modifying drugs because the children have been diagnosed with attention-deficit/hyperactivity disorder (ADHD). Before answering, I always ask if the children can watch a movie all the way through or play video games, and the answer I usually get is that they can do both all day long. In such cases I inform the parents that their children really do not have ADHD. I advise them to wean their kids off some of these highly stimulating activities and spend time reading and discussing books with them instead. Most of the time, when I see the parents again, they tell me that their children's behavior has changed markedly without the drugs. This is not to deny that legitimate cases of organically induced behavioral disturbances exist, but gaining more knowledge about how our brains develop and why they respond as they do to external stimulation will certainly be illuminating. Also in this diverse collection of articles on brain development, we have two pieces examining one of the newest and most intriguing lines of research: the effects of arts education and dancing on learning.

At the other end of the brain-development spectrum, we will gain insight into why some seniors can be quite gullible. We also look at some very controversial issues, such as the use of deep brain stimulation to ameliorate psychiatric conditions and the practice of updating the psychiatric diagnostic manuals as we learn more about the different causes of behavioral disturbances.

As we continue to explore and learn more about this brain of ours, perhaps someday we will actually localize and understand the mind, which all of our sophisticated technology and research have yet to do. Enjoy the adventure!

Articles

The Science of Education

Informing Teaching and
Learning through the Brain Sciences

By Mariale M. Hardiman, Ed.D.,
and Martha Bridge Denckla, M.D.

Mariale Hardiman, Ed.D., is the assistant dean and chair of the department of interdisciplinary studies at the Johns Hopkins University School of Education. She is a former principal in the Baltimore city public school system and author of *Connecting Brain Research with Effective Teaching* (2003).

Martha Bridge Denckla, M.D., is the director of the developmental cognitive neurology department at the Kennedy Krieger Institute and professor of neurology, pediatrics and psychiatry and behavioral sciences at the Johns Hopkins University School of Medicine. Dr. Denckla's research and publications include study of the biological bases for learning disabilities and attention-deficit/hyperactivity disorder (ADHD) in children of normal or above-average intelligence.

The new field of neuroeducation connects neuroscientists who study learning and educators who hope to make use of the research. But building a bridge between these groups will require overcoming some high hurdles: a method needs to be established for translating research findings into educational practices.

As Mariale Hardiman and Martha Bridge Denckla emphasize here, the next generation of educators will need to broaden their approach—focusing not just on teaching math, for example, but also on how math reasoning develops in the brain. Meanwhile, scientists should take the needs and concerns of educators into account as they continue to investigate how we learn. Such crosstalk is already occurring in collaborative efforts focusing on learning, arts and the brain.

RESEARCH SHOWS THAT LEARNING changes the brain. The brain is "plastic"—it makes new cellular connections and strengthens existing ones as we gain and integrate information and skills. In the past decade, the enormous growth in understanding brain plasticity has created an entirely new way to consider how learning and achievement take place in the education of children.

As this knowledge has grown, teachers have increasingly sought to apply it in the classroom. But the link between research lab and school need not be a one-way street—the experiences of educators and students also can suggest questions about learning that neuroscientists should be exploring. Collaboration among educators and cognitive scientists will enrich both fields: Educators can design instructional methods based on research results, and researchers can assess whether these new methods enhance student learning. Such translational research collaborations have the potential to improve teaching and learning and to influence both the practices of school administrators and the policies of boards of education. Neuroeducation—a field establishing how neuroscience research can inform educational practice and vice versa—is taking root.

Yet as educators seek new insights from cognitive research about how people think and learn, they must reconcile extreme opinions in their own field on how—and whether—to apply these findings to the science of teaching. On one end of the spectrum are skeptics who believe that neuroscientific findings have little relevance in the classroom; on the other are people who overstate or misinterpret the results by making claims that certain "brain-based" materials or techniques are sure to improve children's IQs. Our charge to cognitive neuroscientists and educators is to work together to apply compelling, evidence-based findings to teaching and learning, but also to identify misplaced exuberance.

The Science of Learning

Whether or not a teacher understands fundamental concepts derived from basic brain science, such as plasticity, can have a profound effect on how he or she views the learner. Many classroom teachers today, for example, were trained at a time when scientists thought the brain was fixed at birth and changeable only in one direction: degeneration due to aging, injury or disease. Such a misunderstanding of brain anatomy and physiology would limit a teacher's view of the learning capacity of children, especially those who enter the classroom lagging behind their peers. For example, a teacher may think that a fifth-grader who has failed to master basic mathematics skills will always struggle with math because of limited cognitive capacities.

Contrast this view with contemporary knowledge that the brain constantly changes with experience, makes new brain cell connections (synapses), strengthens connections through repeated use and practice, and even produces new cells in certain regions. Imagine how differently a teacher armed with this information would view students' capacity for learning. Knowing that experiences change the brain might encourage this teacher to design targeted remedial lessons. Engaging the student in multiple, creative math-oriented tasks might do more than increase achievement scores: It might actually change brain circuitry.

Cognitive neuroscientists also are providing new insights into the brain's executive functions. For example, we are learning more about the brain's capacity to retain new information in working memory until the tasks that depend on this information are completed. We are also discovering the importance of the cognitive and emotional control people use to arrive at judgments and to make decisions. Valuable contributions to our understanding of these abilities in healthy children have been provided by imaging brain anatomy and connectivity in children with attention-deficit/hyperactivity disorder (ADHD). Findings suggest that ADHD symptoms may represent developmental delay rather than damage in the brain, and that any neural circuitry with such protracted development may be exquisitely sensitive to environmental and experiential influences, which may even alter brain structures.[1] Careful, mindful child-rearing and education are crucial for the development of the brain regions and connections that underlie executive functions.

Research demonstrating the effects of emotions on learning[2] provides another example of how teaching involves not only transmitting information but also crafting classroom climates that promote learning. Teachers may know intuitively that an atmosphere of stress and anxiety impedes children's learning, yet many common practices in classrooms, such as embarrassing a child or making sarcastic rather than constructive comments, can create stressful environments. Teachers who understand that the brain's emotional wiring connects with the prefrontal cortex—the center for higher-order thought—would appreciate the need to provide their students with a positive emotional connection to learning.

Other research highlights the role of motivation in learning and cognition. Studies by Michael Posner, Ph.D., professor emeritus of psychology at the University of Oregon, indicate that children who receive training in a subject that interests them, such as the visual arts, become highly motivated. This motivation sustains their attention, and the result is an improvement in cognition[3] (see Chapter 2, "How Arts Training Improves Attention and Cognition").

Obstacles to Uniting Science and Education

Even as the field of neuroeducation grows, educators will continue to face hurdles. Howard Gardner, the professor of cognition and education at the Harvard Graduate School of Education who developed the theory of multiple intelligences (see Chapter 4), points out that it will be challenging to align the wide-ranging interests of neuroeducators (whom he defines as bench scientists, clinicians, teachers and policy-makers) with the public's notion of effective educational policies and practices.[4] Gardner asserts that with no tradition of applying neuroeducation findings to the practice of teaching, it is difficult to provide benchmarks for good work. We must begin to establish that tradition.

Coherent translation of cognitive neuroscience to education is sparse. We need to translate knowledge about how children learn in ways that are relevant to teachers' work in school settings. Educational policy-makers and administrators focus on the external structures of education, such as standards, data analysis, scheduling, curriculum, school governance and accountability, yet they pay little attention to the learners themselves. And few teacher preparation programs include courses on cognition and learning.

One source of this apparent disconnect is the human tendency to view research findings through the lens of a specific discipline. Neuroeducation, on the other hand, involves examining and synthesizing findings across disciplines. Michèle Mazzocco, director of the Math Skills Development Project at the Kennedy Krieger Institute, emphasizes the need to attend to "both the forest and the trees" when analyzing research relevant to academic performance. Mazzocco, who trained as both an elementary school educator and an experimental psychologist, stresses that the importance of basic cognitive processes sometimes gets lost amid a strong focus only on the result of those processes—student achievement.[5]

For instance, despite their relationship with one another, mathematical ability, performance and achievement involve different cognitive processes, and the study of each facet makes a unique contribution to enhancing student learning. According to Mazzocco, our knowledge of basic cognitive processes, while informative, is not yet sufficiently advanced to provide a solid basis for a specific method of teaching or curriculum. However, teachers are seeking to improve their students' learning and performance right now, and neuroeducators therefore need to determine how best to apply current research findings to improving classroom learning skills.

Setting a New Research Agenda

As neuroeducators seek to address the practical needs of teachers and administrators, they need to conduct more interdisciplinary research to bridge the differences among the methods that scientific and education communities use. Bringing scientists and educators together allows for such intellectual exchange and offers the opportunity to formulate questions that neither group could answer alone.

An example from dyslexia research early in my (Denckla's) career illustrates this point and raises a new one: Input from students about how they learn best, as well as what hinders their learning, can also help direct our investigation. While evaluating children who were having difficulty speaking and reading, I viewed the possible brain basis of their difficulties in terms of a popular notion at that time: that dyslexia is caused, in part, by seeing twisted or reversed symbols (such as confusing the letters *b* and *d*). But one child with dyslexia volunteered that the experiment was "easy and stupid." His explanation indicated that it was the similar *sounds and names* of specific letters, not how they looked, that was confusing. As he explained, his difficulty was in assigning a name or sound to each letter seen alone; *b* and *d* sound too much alike. In contrast, *p* and *q*, another visually similar pair, do not have confusable sounds or names.

Testing a conventional theory, coming up empty and receiving a child's insight steered my research in new a direction: developing an entire line of testing, called the Rapid Automatized Naming Test, which is one of the best predictors of biologically based reading disability. Results from this test, in turn, led neuroimaging researchers to determine where to look in the brain to identify the circuit, or neural connections, normally involved in making automatic the naming of colors, letters and numbers. All of this resulted from a child describing why it was hard for him to learn to read.

To encourage cooperation today, Mary Brabeck, dean of the Steinhardt School of Culture, Education, and Human Development at New York University, suggests establishing a web of collaboration among scientists—including neuroscientists conducting work in medical schools, applied researchers and cognitive scientists working in schools of arts and sciences—and teacher-educators from schools of education.[6] As Brabeck and others have noted, researchers must consider the real needs of educators by visiting schools, engaging in meaningful dialogue and then testing their hypotheses in authentic school settings.

The Learning, Arts, and the Brain educational summit in May 2009, sponsored by the Johns Hopkins University School of Education in collaboration with the Dana Foundation, was an example of such an effort. More than 300 researchers, educators and policy-makers gathered in roundtable groupings to discuss current findings on arts and cognition and to brainstorm ideas for translational research based on educators' questions.

The research reported was preliminary but intriguing, especially the suggestion that skills learned via arts training could carry over to learning in other domains. In addition to Posner's work, Ellen Winner, Ph.D., a professor of psychology at Boston College, and Gottfried Schlaug, M.D., a neurology professor at Beth Israel Deaconess Medical Center and Harvard Medical School, found evidence from brain-imaging studies that music training transfers to the highly related cognitive abilities of sound discrimination and fine motor tasks, a process termed near transfer.[7] Brian Wandell, Ph.D., professor and chair of psychology at Stanford University, described results showing that music training is tightly correlated with phonological awareness—the ability to manipulate speech sounds—a strong predictor of reading fluency, which represents far transfer of cognitive skills.[8]

The research presented at the forum builds on previous studies, including the work of the seven groups of scientists involved in the Dana Arts and Cognition consortium, that show close correlations between arts training and several cognitive abilities. A report from the summit, published in October 2009, reveals rich conversations among scientists and educators that will help shape a research agenda to examine the influence of arts training on creativity and learning.

Another model of collaboration, championed by Kurt Fischer of the Harvard Graduate School of Education and others, is the "research school." Following a medical school model, neuroeducation "residents" would integrate theories of learning and develop practical applications for actual classrooms. Such schools would serve as laboratories for university-based researchers to design and develop studies based on the needs of teachers, test new methods, evaluate interventions and provide teacher-development opportunities.

In this model, a neuroscientist might examine how a specific neurotransmitter, such as dopamine, affects attention; a developmental neurologist might study the delayed structural development of the brain in children with ADHD and compare it with structural abnormalities related to dyslexia; a cognitive scientist might review the neurophysio-logical correlates involved in self-control; educational researchers might assess whether specific types of enriched environments and experiences improved attention for students with ADHD; and teachers might observe instructional interventions that appeared to improve math or reading skills and propose studies to determine how those interventions might affect brain processes.

Effect on Educational Policy and Practice

These types of collaborations would help us start to better align educa-tional practices with evidence from cognitive development studies. For instance, preschool-age children may not be ready for reading instruction, and young adolescents may not be cognitively prepared for the type of conceptual thinking that algebra requires. Studies suggest that connec-tivity of the brain's frontal lobes (which are involved in memory, language, problem-solving, judgment, impulse control, flexibility and social behav-iors) with the neural circuits involved in emotion does not fully mature until about age 32.[9] Cognitive control of the brain's executive function—which is involved in regulating behaviors that are necessary to reaching goals and essential for achieving in school, including the abilities to think abstractly and to form concepts—may not reach maturity until about age 25.[9] Given that these brain processes mature well after students graduate from public schools, what are the implications for the way we teach the relevant subjects? Educators and parents are asking how this information should influence educational practice and wondering who will translate such knowledge from the brain sciences to the educational community.

To provide such information, university research and academic programs, too, must break free from a narrow focus on specific disciplines (such as teaching mathematics) and instead view education through a wider lens that includes the science of learning (such as the development of mathematic reasoning skills). Programs such as the Johns Hopkins University School of Education's new graduate certificate in Mind, Brain, and Teaching and Harvard University's master's degree in Mind, Brain, and Education will produce generations of researchers who are comfortable with an interdisciplinary approach.

Focusing on the science of learning should be as important as accountability for student achievement. Schools' policies and practices must reflect a focus on how children learn, and professionals who conduct learning-related research must view educators as consumers and partners in the work. Education in the 21st century requires a new model for preparing children to become more creative and innovative thinkers and learners. Incorporating multiple perspectives in our study of how children learn can lead us to reimagine and re-create children's learning experiences in our schools.

2

How Arts Training Improves Attention and Cognition

By Michael I. Posner, Ph.D., with Brenda Patoine

Michael I. Posner, Ph.D., is professor emeritus at the University of Oregon and adjunct professor of psychology in psychiatry at the Weill Medical College of Cornell University. Posner, a Dana grantee, is part of the Dana Arts and Cognition Consortium and spoke at the "Learning, Arts, and the Brain" summit at Johns Hopkins University, co-sponsored by the Dana Foundation. He has worked on the anatomy, circuitry, development and genetics of three attentional networks underlying alertness, orienting and voluntary control of thoughts and ideas.

Brenda Patoine is a freelance science writer who has been covering neuroscience for nearly 20 years. She writes regularly for the Dana Foundation (*BrainWork*; *Cerebrum*; *Advances in Brain Research*; *Progress Report in Brain Research*), as well as for the *Annals of Neurology* (NerveCenter news section) and the *NCRR Reporter*, a publication of the National Institutes of Health. Elsewhere on the Web, her work appears on AARP. org (Staying Sharp series), and on alzforum.org and alzinfo.org, two Alzheimer's Web sites.

Does education in the arts transfer to seemingly unrelated cognitive abilities? Researchers are finding evidence that it does. Michael Posner argues that when children find an art form that sustains their interest, the subsequent strengthening of their brains' attention networks can improve cognition more broadly.

IF THERE WERE A SUREFIRE WAY to improve your brain, would you try it? Judging by the abundance of products, programs and pills that claim to offer "cognitive enhancement," many people are lining up for just such quick brain fixes. Recent research offers a possibility with much better, science-based support: that focused training in any of the arts—such as music, dance or theater—strengthens the brain's attention system, which in turn can improve cognition more generally. Furthermore, this strengthening likely helps explain the effects of arts training on the brain and cognitive performance that have been reported in several scientific studies, such as those presented in May 2009 at a neuroeducation summit at Johns Hopkins University (co-sponsored by the Dana Foundation).

We know that the brain has a system of neural pathways dedicated to attention. We know that training these attention networks improves general measures of intelligence. And we can be fairly sure that focusing our attention on learning and performing an art—if we practice frequently and are truly engaged—activates these same attention networks. We therefore would expect focused training in the arts to improve cognition generally.

Some may construe this argument as a bold associative leap, but it's grounded in solid science. The linchpin in this equation is the attention system. Attention plays a crucial role in learning and memory, and its importance in cognitive performance is undisputed. If you *really* want to learn something, pay attention! We all know this intuitively, and plenty of strong scientific data back it up.

The idea that training in the arts improves cognition generally really is not so bold within the context of what we call activity-dependent plasticity, a basic tenet of brain function. It means that the brain changes in response to what you do. Put another way, behavior shapes and sculpts brain networks: What you do in your day-to-day life is reflected in the wiring patterns of your brain and the efficiency of your brain's networks. Perhaps nowhere is this more evident than in your attention networks.[1]

For most of us, if we find an art that "works" for us—that incites our passion and engages us wholeheartedly—and we stick with it, we should notice improvements in other cognitive areas in which attention is important, such as learning and memory, as well as improving cognition in general.

Solid Data Begin to Emerge

If our hypothesis is true, why have scientists been unable to nail down a cause-and-effect relationship between arts education and cognition—for example, "[X] amount of training in art form [Y] leads to a [Z] percent increase in IQ scores"? Such a relationship is difficult to confirm scientifically because there are so many variables at work; scientists have only begun to look at this relationship in a systematic, rigorous fashion.

Early tests of the idea that the arts can boost brainpower focused on the so-called "Mozart effect." A letter published in 1993 in the journal *Nature* held that college students exposed to classical music had improved spatial reasoning skills,[2] which are important to success in math and science. This observation set off a wave of marketing hype that continues to this day. Despite numerous efforts, however, scientists have not reliably replicated the phenomenon. Nonetheless, these studies have involved only brief periods of *exposure* to music, rather than explicit musical training or practice.

More recent attempts to link arts training with general improvements in cognition have relied on a different approach. Researchers have focused on longer periods of engaged participation and practice in arts training rather than simple exposure to music. For example, in 2004, E. Glenn Schellenberg of the University of Toronto at Mississauga published results from a randomized, controlled study showing that the IQ scores of 72 children who were enrolled in a yearlong music training program increased significantly compared with 36 children who received no training and 36 children who took drama lessons. (The IQ scores of children taking drama lessons did not increase, but these children did improve more than the other groups on ratings of selected social skills.)[3]

In a study published in the *Journal of Neuroscience* in March 2009, researchers Ellen Winner of Boston College, Gottfried Schlaug of Harvard University and their colleagues at McGill University used neuroimaging scans to examine brain changes in young children who underwent a four-year-long music training program, compared with a control group of children who did not receive music training.[4] In the first round of testing, after 15 months, the researchers found structural changes in brain circuits involved in music processing in the children who received training. They did not find the same changes in the control group. The scientists also found improvements in musically relevant motor and auditory skills, a phenomenon called near transfer. In this case, the improvements did not transfer to measures of cognition less related to music—termed far transfer. We do not know why far transfer to IQ, for example was found in the Schellenberg study and not in this one.

Taken as a whole, the findings to date tell us that music training can indeed change brain circuitry and, in at least some circumstances, can improve general cognition. But they leave unsettled the question of under what circumstances training in one cognitive area reliably transfers to improvements in other cognitive skills. From our perspective, the key to transfer is diligence: Practicing for long periods of time and in an absorbed way can cause changes in more than the specific brain network related to the skill. Sustained focus can also produce stronger and more efficient attention networks, and these key networks in turn affect cognitive skills more generally.

Practicing a skill, either in the arts or in other areas, builds a rich repertoire of information related to the skill. Scientists conducting neuroimaging studies of many human tasks have identified networks of widely scattered neural structures that act together to perform a given skill, which may involve sensory, motor, attentional, emotional and language processes. The arts are no exception: Specific brain networks

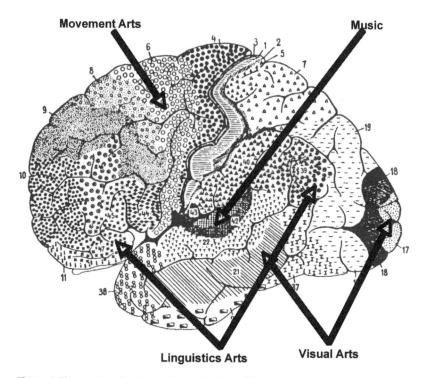

Figure 1 The practice of various art forms involves different sensory and motor areas in the brain. (Courtesy of M. Posner.)

underlie specific art forms, as illustrated in Figure 1. As we practice a task, its underlying network becomes more efficient, and connections among brain areas that perform different aspects of the task become more tightly integrated.

This process is analogous to an orchestra playing a symphony. The music that results from the integration of orchestral sections is likely to sound more fluid the hundredth time they play a piece than the first time.

Training Attention Networks

A large body of scientific evidence shows that repeated activation of the brain's attention networks increases their efficiency. Neuroimaging studies have also proved that the following specialized neural networks underlie various aspects of attention[1] (see Figure 2):

- the *alerting* network, which enables the brain to achieve and maintain an alert state;
- the *orienting* network, which keeps the brain attuned to external events in our environment;
- the *executive attention* network, which helps us control our emotions and choose among conflicting thoughts in order to focus on goals over long periods of time.

I have been particularly interested in the executive attention network. Executive attention skills, especially the abilities to control emotions and to focus thoughts (sometimes called cognitive control), are critical aspects of social and academic success throughout childhood. Empathy toward others, the ability to control reward-motivated impulses and even control of the propensity to cheat or lie have been linked scientifically to aspects of executive attention.[5] Researchers also have shown that measures of this network's efficiency are related to school performance.[6]

Given the importance of the executive attention network, my colleagues and I wondered what might improve its efficiency. To find out, we adapted a series of exercises, originally designed to train monkeys for space travel, to investigate the effects of attention-training exercises in 4- to 6-year-old children. We randomly assigned the children to either a control condition (which involved watching and responding to interactive videos) or training on joystick-operated computer exercises designed to engage attention networks through motivation and reward. After the children who did the computer exercises participated in five days of training for about 30 minutes per day, we placed noninvasive electrodes on the children's scalp to look at their brain activity; we found evidence of increased efficiency in the executive attention network. The experimental group's network performance, in contrast to the control group's, resembled performance in adults. Importantly, this improvement transferred to higher scores on IQ tests designed for young children.

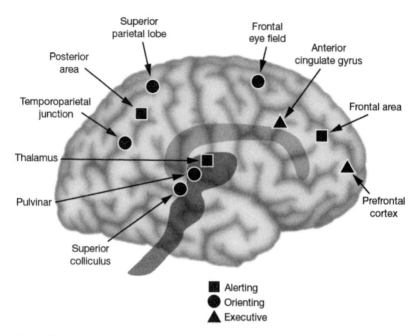

Figure 2 Brain networks that underlie different aspects of attention include the alerting network, the orienting network and the executive attention network. Arts learning may contribute to improved cognition by improving the efficiency of the executive attention network. (Courtesy of M. Posner.)

These data suggest that increasing the efficiency of the executive attention network also improves general cognition as measured by IQ.[7] M. Rosario Rueda of the University of Granada, Spain, and colleagues subsequently replicated this key finding in an as yet unpublished study of Spanish children. Rueda found that attention training improved the children's abilities to delay reward, and the improvements persisted for at least two months after training.

In recent years, various approaches to training children to pay attention have been carried out in many different settings. The results show that tasks specifically designed to exercise the underlying networks can indeed improve attention, and that this kind of training can translate to better general cognition. In one of the strongest studies to support this finding, measures of cognitive control significantly improved in preschoolers enrolled in a yearlong training program that incorporated different activities designed to sharpen executive functions.[8] We expect that this training will positively affect the children's future academic performance, but this remains to be shown.

For many children, interest in a particular art form leads to sustained attention when practicing that art form. Moreover, engaging in art often involves resolving conflicts among competing possible responses, such as when choosing the correct note to play at a given moment. The ability to resolve conflict among competing responses is also a crucial aspect of attention training. For example, if you are to respond to a target arrow by pressing a key in the direction in which the arrowhead points, the addition of surrounding arrows pointing in the opposite direction will increase your reaction time and activate parts of the executive attention network[8]. We expect, therefore, that arts training should exercise the executive attention network and, therefore, also should improve cognition generally.

One Size Doesn't Fit All

It seems unlikely that training in the arts will *always* improve general cognition, however, since so many factors are at play. No single art form is interesting to all people, and some people may never warm up to any type of art. Individual differences in relevant brain networks, which are probably genetically influenced to some degree, help explain this variability in both appreciation of and ability to create art. For example, one person may have an auditory system that easily discriminates between tones and a motor system optimized for fine finger control, which may predispose her to playing a musical instrument. Someone with agility, coordination and a good ability to imitate motions of others, on the other hand, might naturally gravitate toward dance or sports. These differences may also help explain why people are passionate about one type of art but not others.

The efficacy of arts training also depends on a child's temperament or personality. For example, openness, which affects behavior, may be a prerequisite to effective training, and may in part be genetically derived.

We have found, for instance, that a gene that regulates the transmission of the chemical dopamine from one brain cell to another appears to modulate children's openness to parental influence. Our studies show that children with one form of this gene (the dopamine-4 receptor gene) show abnormally high sensation-seeking behavior if their parents show poor parenting skills, but not if their parents show good parenting skills.[9,10]

An increasing body of evidence indicates that the brain's attention networks are also under some degree of genetic control. For example, certain genes seem to modulate an individual's ability to perform attention-related tasks, such as quickly responding to a warning signal or shifting attention from one external event to another. These genetic influences underscore individual differences in responses to training, and they may explain contradictory results in scientific studies investigating the links between arts training and cognition.

Apart from these caveats, exposure to the "right" art form can fully engage children's attention and can be highly rewarding for them. They may get so involved in learning the art that they lose track of time or even "lose themselves" while practicing it. I believe that few other school subjects can produce such strong and sustained attention that is at once rewarding and motivating. That is why arts training is particularly appealing as a potential means for improving cognition. Other engaging subjects might be useful as well, but the arts may be unique in that so many children have a strong interest in them.

With advances in neuroscience that are providing important new tools for studying cognition, it is important for researchers to work with educators to design and carry out studies that build upon the findings that arts training provides near-transfer effects, and determine whether this training also results in—and causes—far-transfer cognitive benefits. As we have seen, recent studies have transcended the failed paradigm of simply exposing people to the arts, and now concentrate on the effects of arts training over months and years. We need more studies like these to determine whether, beyond strong correlation, causation occurs. Arts training may influence cognition through other brain processes as well. Because arts training strengthens the brain network related to the art being practiced, other tasks that rely on the same brain circuitry or pieces of it presumably would be affected. For example, if music training influences the auditory system, we might also expect to see improvement in nonmusical tasks involving pitch. In fact, Brian Wandell and his colleagues at Stanford University recently demonstrated that children who train in music or the visual arts showed improved phonological awareness, the ability to manipulate speech sounds, which is strongly tied to reading fluency. Moreover, the more music training they had, the better their reading fluency.[11]

In addition, parts of the music network lie adjacent to brain areas involved in processing numbers, which might explain anecdotal reports of improvements in mathematics after music training. For instance, Elizabeth Spelke of Harvard University has found that school-age children engaged in intensive music training had improved performance in abstract geometry tasks.[12] Wandell and his team also reported preliminary data connecting experience in the visual arts with children's math calculation abilities.[13] Future studies will need to examine these possibilities in more detail.

Another interesting aspect of the performing arts is that artists often prepare for their work by consciously entering a state of mind that they believe will elevate their performance, for example, via deep breathing, picturing the moment or other meditative techniques. Yi-Yuan Tang, a visiting professor at the University of Oregon from Dalian Medical University in China, recently reported that some forms of meditation can produce changes in the connection between the brain and the parasympathetic branch of the autonomic nervous system and, after just a few days of training, can lead to improvements in the same aspects of executive attention that are trained by specifically exercising this network.[14] This "attention state" also correlates with improved mood and resistance to stress. Our data suggest that meditation may contribute to generalized cognitive improvements in those who practice it.

The growing body of scientific work that suggests arts training can improve cognitive function—including our view, which identifies stronger attention networks as the mechanism—opens a new avenue of study for cognitive researchers. The new research findings also give parents and educators one more reason to encourage young people to find an art form they love and to pursue it with passion. Continuing research in this area can also help inform ongoing debates about the value of arts education, which has important policy implications given budgetary pressures to cut arts programs from school curricula.

From our perspective, it is increasingly clear that with enough focused attention, training in the arts likely yields cognitive benefits that go beyond "art for art's sake." Or, to put it another way, the art form that you truly love to learn may also lead to improvements in other brain functions.

3

What Can Dance Teach Us About Learning?

By Scott T. Grafton, M.D.

Scott T. Grafton, M.D., is a professor of psychology and director of the Brain Imaging Center at the University of California, Santa Barbara. From his background in clinical neurology Grafton became interested in how our brains allow us to accomplish complex physical goals. Grafton is a leader in using functional brain imaging to reveal how the brain changes as a result of practice.

We might begin to learn a dance step when someone describes it to us, but we learn it better when we physically perform the steps as we observe and imitate an instructor doing them. Scott Grafton's research sheds light on the brain's action observation network, which fires up both when we perform an action and when we watch someone else perform it. Dr. Grafton contends that his and others' findings highlight the importance of including physical learning in the classroom, to stimulate creativity, increase motivation and bolster social intelligence.

WE USE MANY MENTAL AND PHYSICAL STRATEGIES, some more effectively than others, to learn new skills. Consider the challenge of learning a new dance step—whether something silly, such as the Hokey Pokey, or elaborate, such as a series of flamenco steps or a hip-hop "pop." Verbal descriptions of how, where and when to move our bodies can work pretty well if the dance isn't too complicated. As movements become more complex, however, our capacity to follow verbal instruction decreases. Words are too slow, general and serial to encapsulate all the details of a precise move.

One alternative to verbal instruction is to follow spatial guides—for example, by stepping on patterns traced on the floor. This old-fashioned method of dance instruction forms the basis of extremely popular computer dancing games such as *Dance Dance Revolution* and music games including *Guitar Hero* and *Rock Band*. To gain points, players must hit sequences of spatial targets at the correct times (with a song acting as a metronome) and with the correct body parts (feet, fingers or hands).

Observational Learning

Another way to learn is to observe and imitate a physical model. For example, thousands of online instructional videos use physical models to show us how to dance or how to play musical instruments. Through trial and error, learners refine their skills. Research shows that teaching via a physical model improves learning more than verbal descriptions or spatial guides. We also know that learners acquire a motor skill more rapidly and precisely if the model is a real person placed alongside the learner.[1] A real dance instructor, for instance, can adapt a lesson to individual strengths and weaknesses.

While words can help us start to learn a dance, ultimately we need to form nonverbal knowledge—what coaches incorrectly refer to as "muscle memory"—in order to dance anything complicated. Ample evidence shows that brain networks, not muscles, store memories of movements such as dance steps. These memories, which are called motor memories because they are specific to movement, are fundamentally different from memories of verbal descriptions of these same actions.

My research team, which is interested in how people create motor memories, studies the interplay between the brain networks we use to understand others' actions and the networks we use to generate movements. Observational learning is the *process* of modeling others' actions by observing their movements. This process has implications as far beyond the realm of dance as classroom education. We believe that neuroscientists' insights about observational learning are highly relevant to the growing field of neuroeducation and should influence decisions about what is taught in K–12 classrooms.

There are two questions critical to understanding observational learning: Does the brain use similar circuits both to recognize and to perform an action? And, if so, does this overlap constitute the neural basis for observational learning? Dance is a solid experimental medium for testing these questions, and our tests have yielded evidence that the brain has a shared network for observing and doing. This network allows us to simulate action and is thus a powerful learning engine.

A Familiar Dance Lights Up the Brain

Watching a video of Michael Jackson dancing fires up a widespread network in the cortex across the brain's hemispheres.[2] We have named this circuitry the *action observation network* (AON). To observe it we use magnetic resonance imaging (MRI), which creates images that reflect changes in brain blood flow and indicate which areas are active during a task. Strikingly, the AON includes many brain areas that are also active during actual movement. This fact contradicts classic descriptions that divide the brain into sensory and motor areas. The AON is a storehouse of physical knowledge—of the self and of other objects—that can be used both to understand and to plan action.

Discovery of the AON has revitalized interest in motor simulation theory, which holds that we use our own motor memories to figure out what other people are doing. When we watch a video of a dancer, motor areas of the brain might activate automatically and unconsciously—even though our bodies are not actually moving—to find familiar patterns that we can use to interpret what we are watching. In other words, some sort of resonance takes place between the circuits for observing and for doing. If this is true, the AON should be more active when we observe actions that are physically familiar than it is when we observe unfamiliar actions.

This theory is hard to test with everyday actions; it is difficult to find someone who is unfamiliar with picking up a coffee mug or walking across a room. In contrast, people differ significantly in their levels of dance experience and competence. For example, while many of us are familiar with the moonwalk, few of us can actually create this movement. Thus, we might expect to see more brain activity in someone watching a familiar dance than in someone watching the moonwalk.

In an innovative paper published in 2005, Beatriz Calvo-Merino and colleagues at University College London used functional MRI (fMRI) to compare brain activity in two groups of trained dancers—one skilled in ballet and the other in capoeira, an Afro-Brazilian dance based on martial arts movements. Both groups were exceptionally skilled at dance, but each knew a set of profoundly different dance movements. The study found greater activity in the AON when dancers watched videos of familiar dances than when they watched videos of unfamiliar dances. This result is consistent with the simulation hypothesis.[3]

Each troupe's dances were both visually and physically unfamiliar to those of the other troupe. This was a research-design weakness because it raised the possibility that the AON became active due to visual familiarity rather than physical experience. Calvo-Merino's team addressed this question in a clever follow-up study in 2006. Male and female ballet dancers observed videos of ballet movements that only men or only women perform. The dancers were visually familiar with all movements but physically familiar only with those in which they were trained. Again, AON activity was greater when dancers observed movements with which they were physically familiar.[4]

Calvo-Merino's results suggest that dancers might perceive the world differently because they have a special capacity to simulate what they observe. This theory does not imply that we can understand actions only if we have already performed them. Were that the case, the moonwalk would be uninterpretable. The theory suggests only that prior experience amplifies the ability to simulate others' actions.

Although researchers have not yet tested this idea directly with dancers, recent studies of athletes provide some preliminary support for it. Imagine watching a skilled basketball player shoot free throws. Your task is to judge, as quickly as possible, whether or not each shot will go in the basket. In 2008, in an analogous approach using videos, Salvatore Aglioti and colleagues at the University of Rome compared the judgments of professional basketball players, coaches and fans. Players were far better than fans at quickly and accurately predicting whether a free throw would go in. The athletes unconsciously detected the shooter's subtle movements, such as the angle of his ankle and wrist as he released the ball.[5] Coaches, who averaged about seven years of playing experience, were better than fans but not as good as players. These results suggest that extensive practice—*embodied knowledge*—changes our ability not just to execute, but also to observe.

Rapid Training Enhances Effects

Ballet and capoeira dancers train for years. Can we measure changes in the AON after shorter training periods? If the answer were yes, such a finding would suggest that the AON is also involved in the kind of observational learning that we use at a moment's notice to pick up new skills. To test this, Emily Cross from my laboratory trained a troupe of modern dancers on a new dance piece over ten weeks and tracked changes in AON activity.[6] While their brains were scanned using fMRI, the dancers watched short video segments of their instructor performing one dance piece that they rehearsed daily and a second piece that they had seen but not practiced. As they gained physical experience in the dance they practiced and watched the video of their instructor, activity in the AON increased. Moreover, while watching the video segment of the rehearsed dance, the better a dancer thought he or she could execute a part of the dance, the greater his or her AON activity. This finding provides new evidence that links physical competence and simulation within the AON.

We next investigated whether the AON is modified only by physical practice or also by observing an action. We trained undergraduates with limited dance experience on a set of challenging dances using a variation of *Dance Dance Revolution*. Each dance consisted of a sequence of stepping patterns associated with a specific song; arrows provided foot-placement cues.[7] Participants learned some dances through physical practice and other dances solely by observing the instructions. Follow-up tests showed that the benefit from learning by observing was never as strong as advantages derived from physical practice.

These participants also were scanned while they watched segments of the different dances. Activity in the AON was greater when they watched dances that they had practiced physically compared with dances to which they had not been exposed; this is consistent with findings from the other fMRI dance studies. But this experiment revealed something new: *Dances learned by observation alone triggered AON activity, but dances learned via physical practice stimulated greater activity.* We also found that adding a physical model (including an image of a dancer performing the particular dance piece) augmented brain activity compared with a video that used only symbolic instructions (arrows).[8] These findings led us to conclude that the AON is a general-purpose observational learning network for simulating actions, and it is cued by both physical and symbolic models.

Action Observation as an Engine for Learning

The AON experiments provide glimpses into a brain system that is exquisitely tuned to learn and to understand physical knowledge. Such insights from cognitive neuroscience help to identify learning principles important to education policy and practices; they form the very basis of the new (and growing) academic field of neuroeducation. Our society places great emphasis on academic success within a narrow set of areas (math, reading or higher IQ). This makes it tempting to apply a narrow translational framework whereby the value of dance, music and other physical arts depends only on the degree to which these pursuits make a student more successful in the desirable academic areas.

Within this narrow framework, some evidence suggests that training children in music modestly improves their IQ scores.[9] Investigators, including me, who are participating in a Dana Foundation-supported research consortium, are identifying some of the many factors that likely contribute to these modest improvements. For instance, Michael Posner at the University of Oregon is showing that children who are interested in learning music focus their attention on it. They can then apply their improved attention skills to learning to read, write, or solve math problems.[10]

A more challenging question is whether learning in one area transfers to others. For example, Elizabeth Spelke at Harvard University is testing the hypothesis that improved understanding of the complex spatial and temporal patterns of music could generalize to improved abilities in abstract spatial reasoning, which underlies some mathematics—especially geometry. In this model of studying whether music training transfers to improved abilities in spatial reasoning, though, there is no examination of whether the physical skills involved in the training might contribute to the potential effects.

Scientists test whether arts learning transfers to other cognitive areas using behavioral measurements in children before and after they receive arts training. We know a lot about brain areas activated in both pursuit of the arts and in cognition. Two important tasks are to figure out how to use the growing wealth of functional imaging data to extend ideas about crossover effects and to see if this new information might further influence education theory. For example, brain areas activated during mathematical reasoning overlap strikingly with those active when watching dancers (for example, components of the AON). Can we then conclude that learning to dance well translates to mathematical pursuits because the same brain area is involved in both? Though plausible, this idea lacks sufficient behavioral evidence to merit a change in education practice. It is likely an oversimplification that doesn't accommodate for the degree to which specific brain areas are actually part of much larger and more complex neural networks. This example reminds us that the pathway from neuro-science insight to educational practice will be more complicated than a linear translation of principles.

Conventional K–12 education practitioners emphasize reading, writing and mathematics, which supposedly will produce progressive parallel development in cognitive capacities such as reasoning, abstraction and semantic knowledge. School administrators add physical education to the curriculum mostly for its health benefits. At the same time, they are cutting courses focused on material or physical knowledge, such as machine or wood shop, electronics, music, dance and theater arts. This approach belies an assumption that anyone with sufficient cognitive abili-ties can gain material or physical knowledge if they just put their mind (or hands) to it; as long as children are smart in the head, the hands will follow. This false dichotomy between knowing (as defined by cognitive models) and doing dominates contemporary education policy and practice.

Physical Mastery, Motivation and Social Intelligence

Our AON research suggests three ways that learning in the arts—and in physical skills more generally—remains vital to educational practice.

First, studies of the AON and observational learning remind us that much knowledge comes from the overwhelming challenges of mastering the material world, be it changing a flat tire or playing a song on a musical instrument. Physical knowledge need not be subordinate to cognitive training for a child to receive the best education. Experiential knowledge is essential for creating great surgeons and truck drivers alike. It drives creativity and innovation in the sciences, as anyone who has ever built a novel measurement device, developed a new laboratory assay or handled a sample of any kind can attest. The bulk of this knowledge stems from observational and imitative learning, which the AON facilitates. Success in solving problems in the real world, not the virtual or symbolic world, gives most people their deepest joys; ultimately, we researchers would like to understand how.[11]

The second benefit of teaching arts and physical knowledge is based on harnessing passion. The hunger to learn these skills is a source of profound motivation that can spread to all aspects of a learner's life and augment performance generally. Teachers see this every day. We are only just beginning to understand how the brain creates positive motivation. Our research also has focused on defining brain systems that are driven by passionate interests, physical skills and how they interact in the AON.

For example, in one experiment, we briefly flashed a string of letters on a computer screen and asked participants to decide if the letters constituted a word (e.g., *world*) or a nonword (e.g., *owrdl*). In a sneaky manipulation, we also flashed a word that described either an activity the participant was passionate about, such as *guitar*, or an activity about which he or she did not feel strongly. This word appeared just before the word/ nonword challenge, and it happened so quickly that the person was not consciously aware of it. Nevertheless, participants were faster and more accurate at the word/nonword task when their "passion" word appeared.[12] We also tested this effect in patients with Parkinson's disease and found that "treatment" with these subliminal cues can improve performance in multiple tasks for about 20 minutes. Furthermore, brain scans showed that the AON shapes and modifies the interaction between word priming and the subsequent word/nonword recognition task. These results show that *passions can be a source of motivation that spreads to a broad range of cognitive challenges.* A skillful teacher thus might be able to use the arts to harness students' enthusiasm and help spread it to other subjects—a point Michael Posner makes in Chapter 2, "How Arts Training Improves Attention and Cognition."

Third, studies in the arts amplify learning that supports social intelligence. Specifically, the emotional scaffolding that supports empathy and perspective is linked in part to how we perceive and interpret others' actions. In recent work, we have been able to show direct involvement of the AON not only in what people are doing, but also in how they feel as they do something. When you watch someone walk down the street, your AON is highly tuned in to his or her body language—signals that suggest whether the person is happy or sad, for example. This initial research inspires future studies that should be able to determine the degree to which the AON can change as emotional intelligence develops, as well as the potential for education in the arts to accelerate this process.

Could Action Observation Help Patients with Brain Injuries?

Imaging studies of the action observation network (AON) have helped scientists identify a powerful learning system in the brain that links perceived actions and self-generated behavior. These findings are generating excitement in the field of rehabilitation medicine because they suggest new approaches to help train people trying to regain motor function after brain injury from stroke or trauma. Observing others' actions, the theory goes, might activate still-healthy brain circuits—including the AON—and accelerate recovery. There are two potential limitations, however.

First, it would be naive to assume that all observations of actions in the physical world can directly transfer to performing a similar action without some trial-and-error learning. While observational learning supported by the AON probably provides a rough template for movement, real physical practice ultimately is unlikely to have a substitute. Therefore, a patient must have enough undamaged motor pathways from the brain to the spinal cord in order to generate some level of action to relearn given movements.

Second, the AON almost certainly depends on a large library of motor programs that we acquire throughout brain and body development, and these programs might be irreplaceable. We generally underestimate how much training occurs as a child develops into an adult. Consider a 14-month-old toddler learning to walk. He takes about 2,000 steps (almost half a mile) and 15 falls a day in a whirlwind of fearless trial-and-error exploration.[13] A few weeks of observational learning in an adult is unlikely to replicate the effects of many years of training.

Integrating these ideas, it is reasonable to propose that patients with brain injuries will make the greatest recovery when the training environment employs their existing movements and requires problem-solving that uses observational learning over a long period of time.

4

A Debate on "Multiple Intelligences"

A Cerebrum Classic*

By Howard Gardner, Ph.D., and James Traub

Howard Gardner is the Hobbs Professor of Cognition and Education at the Harvard Graduate School of Education. A leading thinker about education and human development, he has studied and written extensively about intelligence, creativity, leadership and professional ethics. Gardner's recent books include *Changing Minds, The Development and Education of the Mind* and *Multiple Intelligences: New Horizons*. His latest book, *Five Minds for the Future* (Harvard Business School Press), was published in 2007, and in 2009 he co-edited *Multiple Intelligences Around the World* (Jossey-Bass).

James Traub is a contributing writer for *The New York Times Magazine*, where he has worked since 1998. He has written extensively about international affairs and especially the United Nations, as well as national politics, urban affairs and education. His most recent book is *The Freedom Agenda: Why America Must Spread Democracy (Just Not the Way George Bush Did)* (Farrar, Straus and Giroux, 2008). In 2006 he published *The Best Intentions: Kofi Annan and the UN in the Era of American World Power*. He is a member of the Council on Foreign Relations and speaks widely on international affairs.

---

* From Volume 1, Number 2, Fall 1999

Authors' Notes to Readers in 2010

Howard Gardner

I first introduced the theory of multiple intelligences (MI) more than 25 years ago, in 1983. Developed primarily as a psychological theory of how the mind has evolved and how it is organized today, the theory received its greatest attention and acclaim in the field of education. A decade ago, James Traub, a noted journalist, and I engaged in a written debate about the merits and applications of the theory. Inasmuch as we are both primarily writers, this format proved appropriate and allowed for more thoughtful interaction than is usually the case in an oral debate.

Since then, Traub has moved on to other areas, as have I. To our surprise, however, the interest in MI theory has persisted and, indeed, has increased in some quarters. Also, my colleagues and I have revised and extended the theory in light of additional research and reflection. Readers interested in the most up-to-date account of the theory should read the forthcoming "The Theory of Multiple Intelligences" by Katie Davis, Scott Seider, Joanna Christodoulou and me, in *The Cambridge Handbook of Intelligence*, edited by Robert J. Sternberg and James C. Kaufman.

MI theory has also been the focus of three books, each of which includes critical comments and my responses:

- *Howard Gardner Under Fire*, edited by Jeffrey A. Schaler (Open Court Publishing, 2006).
- *Multiple Intelligences Around the World*, edited by Jie-Qi Chen, Seana Moran and Howard Gardner (Jossey-Bass, 2009).
- *MI at 25*, edited by Branton Shearer (Teachers College Press, 2009).

James Traub

As Howard Gardner notes, both of us have moved on to other subjects. I'm not prepared to assess the standing of multiple intelligences in the world of psychology, where its scientific merits will be judged. My chief interest was always the way in which MI found a home in a very different world—that of pedagogy and school reform doctrine. The idea that there are different ways of learning does indeed seem to have become a fixed part of education's discussion—and one for which Gardner can take much credit. I remain concerned, however, that this truth—or, rather, this premise—not overshadow what strikes me as a much more important one, which is that children who are at risk of failure in school, and thus in all likelihood at risk of failure later in life, must attain mastery of basic skills through the most simple and direct means possible. Everything else is a luxury.

The scientific study of intelligence began more than a century ago with Sir Francis Galton, an aristocratic polymath whose own IQ has been estimated posthumously at 200. A cousin of Charles Darwin, Galton became fascinated with inherited abilities. In 1869, with the publication of Hereditary Genius, *Galton advanced the thesis that intelligence is an inherited human characteristic. He also brought into play the statistical approaches to measuring IQ that have dominated psychometrics—the science of measuring mental attributes—ever since.*

At the beginning of [the 20th] century, a Frenchman, Albert Binet, director of the Sorbonne psychology lab, introduced widespread testing of schoolchildren in Paris to ascertain their intellectual abilities. Binet and his colleague, psychiatrist Theodore Simon, gave us the concept of a "mental age" and the nascent industry of intelligence testing. To round out the story, a German psychologist, William Stern, suggested that an "Intelligence Quotient" could be calculated by dividing mental age by chronological age (and multiplying the result by 100 to get round numbers). To emphasize the supposed unity of the trait being measured by intelligence tests, psychologists came to refer to it as "G" for general intelligence.

Summing up these early years of psychometrics, Jack Fincher, in Human Intelligence *(G.P. Putnam, 1976), wrote: "The siren song of an intelligence that was stable, distinct, and independent caught the public ear as few seductive sounds from science have before or since."*

IQ testing has exerted its influence in the United States at least since World War I. But in recent decades, critics of IQ testing and of the concept of a "stable, distinct, and independent" trait called intelligence have proliferated in psychology and education. They charged that the field uses instruments that discriminate against (to name a few) minorities, the exceptionally bright, the creative and the poor. None of those challenges has had the impact of Howard Gardner's 1983 book Frames of Mind, *which introduced the concept of "multiple intelligences"—relatively autonomous faculties including the linguistic, logical-mathematical and musical. Although Gardner, a professor of developmental psychology at Harvard University, put forward his work as a psychological theory, not a platform for school reform, its impact on schools in America—indeed worldwide—has been great. The outpouring of books, articles, conferences, courses and curricula invoking the concept of multiple intelligences might bring to mind Fincher's melodious phrase, "caught the public ear as few seductive sounds from science have before or since."*

But Gardner, too, has his critics. James Traub, an education writer, recently challenged Gardner's ideas and their effect on education in an article entitled "Multiple Intelligences Disorder" in The New Republic *(October 26, 1998). Traub is the author of* City on A Hill: Testing the American Dream at City College *(Addison-Wesley, 1994), a* New York Times *Notable Book of the Year and winner of the 1994 Sidney Hillman Foundation Book Award. He has written about education for* The New Yorker, Harper's *and other magazines and is now a contributing writer for* The New York Times Magazine.

Cerebrum *invited Gardner and Traub to debate the concept of multiple intelligences in these pages. Here they debate three major questions: "What is the evidence for the theory of multiple intelligences?" "What has been the impact on education?" and "What is the potential of this idea for the future?"*

Gardner and Traub addressed these questions in initial statements. They then exchanged statements and each wrote a rejoinder. Finally, they exchanged rejoinders so that each could write a short closing statement. We begin with Gardner.

Gardner's Opening Statement

What Is the Evidence for the Theory of Multiple Intelligences?

For nearly a century most psychologists have embraced one view of intelligence: Individuals are born with more or less intellectual potential (IQ); this potential is heavily influenced by heredity and difficult to alter; and experts in measurement can determine your intelligence early in life, currently from paper-and-pencil measures, perhaps eventually from examining the brain in action or even scrutinizing your genome. Richard

Herrnstein and Charles Murray defended this position in their controversial book *The Bell Curve* (1994).

Recently, criticism of this conventional wisdom has mounted. Biologists ask if speaking of a single entity called "intelligence" is coherent and question the validity of measures used to estimate heritability of a trait in humans, who, unlike plants or animals, are not conceived and bred under controlled conditions.

A research psychologist by training, I began to doubt the standard view as a result of my work with both normal and talented children and with brain-damaged adults. I was struck that individuals can be strong (or weak) in certain skills, but that strength cannot predict skill in other areas. If a person is strong in telling stories, solving mathematical problems, navigating unfamiliar terrains, tracing the transformations of a fugal theme or understanding the motivations of others, one does not know if comparable strengths (or weaknesses) will be found in other areas. That intuition lies behind the "theory of multiple intelligences" (MI theory).

While other scholars have had this intuition, most have tried to support it by an examination of the correlations (or, rather, the lack thereof) of scores on paper-and-pencil-style tests of these skills. I approach the scientific challenge differently. I define intelligences as biological potentials to process information in certain kinds of ways to solve problems or fashion products valued in one or more cultures. To see if there is evidence for a possible intelligence, I examine eight separate criteria, drawn principally from biology, anthropology and psychology. These criteria range from the existence of special populations that are strikingly strong or weak in an area (e.g., prodigies, savants) to the existence of symbol systems (e.g., maps, dance notation) that capture specific forms of information.

A comment on two biologically oriented criteria. One is "a distinct evolutionary history." Existence of a specific intelligence becomes more plausible if one can locate its evolutionary antecedents, including capacities (like bird song or primate social organization) that humans share with other organisms. Specific capacities may operate in isolation in other species but have become yoked in human beings. For example, discrete aspects of musical intelligence may appear in several species but be joined in human beings. In *The Prehistory of the Mind*, Steven Mithen describes how the several intelligences may have evolved one after the other.

Another critical criterion is "potential isolation by brain damage." Scientists used to think that the brain had the same potential throughout, but now they agree that different regions of the brain serve different functions. While avoiding a simple-minded phrenology, it is reasonable to speak of the middle frontal and temporal regions of the left hemisphere as language areas, and the posterior regions of the right hemisphere as spatial areas (in right-handed adults). When one can point to regions dedicated to certain processes, and when injury to these regions compromises those processes, one has persuasive evidence of a separate intelligence.

Based on these criteria, I originally identified seven intelligences. Each can be exemplified by a vocation that draws heavily on that intelligence. Thus the poet is strong in linguistic intelligence, the composer in musical intelligence and the salesperson in interpersonal intelligence.

This way of conceptualizing intelligence has two corollaries. First, all human beings possess each of these intelligences; this, in fact, constitutes a definition of being human, cognitively speaking. (Rats may have more spatial intelligence, computers may have more logical-mathematical intelligence, but neither has intrapersonal intelligence.) Second, no two individuals, not even identical twins, possess exactly the same mosaic of intelligences because each of us has different experiences.

My reading of the scientific literature following the 1983 publication of *Frames of Mind* provides strong confirmation of the approach I took and the intelligences I identified. With this new evidence, I could identify an eighth (naturalist) intelligence, and a possible ninth (existential) intelligence. Suggestive findings about a possible relationship between musical and spatial processing may require reformulation of those two intelligences. New in vivo methods of examining the human brain enable us to specify capacities more accurately. Finally, brain studies reveal two important correctives to my original approach: first, the greater idiosyncrasies across individual persons; second, the involvement of many neural regions in nearly all non-elementary skills. Were I rewriting *Frames of Mind*, I would look for evidence that each intelligence involves specific (but not necessarily identical) regions of the brain in the majority of individuals examined.

What Has Been the Impact on Education?

I in no way anticipated the enormous interest in MI theory on the part of educators. The interest initially came from those involved with special populations (children with learning problems, "gifted" children) and those working with young children. More recently, interest has extended to middle schools, secondary schools, colleges, museums and even the workplace. This interest has been notably sustained. Whatever its scientific merits, clearly MI theory has struck a responsive chord in educators in the United States and abroad. It is important to indicate that this interest has developed with little push from me. I wrote initially as a psychologist, not as an educator. I have neither endorsed any program nor had any commercial interest in any applications of my theory. The response to MI theory has arisen, without prodding, from the worlds of education and training.

Given this almost laissez-faire atmosphere, it is not surprising that many applications of MI theory to education have been proposed. (In a new book, *Intelligence Reframed*, I list about 500 books and articles on MI.) Indeed, the theory has functioned as an ink blot, with individuals reading into it their own hopes or anxieties. There is no way for me to track all of these applications. Recently, in her SUMIT (Schools Using Multiple Intelligence Theory) study, my colleague Mindy Kornhaber has systematically studied 41 schools using multiple intelligences theory.* Her survey reveals that a majority of these schools indicate improvements in both scholastic (test-score) and other measures (parent involvement, student discipline). The theory has proved especially productive for students with learning difficulties.

Applications coming to my attention have been largely benign. MI theory has been used as a rationale for including a wider range of cognitive

* For current details, see http://www.pz.harvard.edu/Research/SUMIT.htm.

activities in the school day (and after school), setting up "learning centers" or "flow rooms" where students pursue their own interests, encouraging students to carry out rich projects and broadening the basis on which student understanding is assessed.

On occasion, however, the applications have been pernicious. Most upsetting was a curriculum in Australia that labeled children from different ethnic groups as having (or lacking) specific intelligences. Not only does this statement have no scientific basis; it offended my (and presumably most other people's) sensibilities. I went on television to dissociate myself from the curriculum and was relieved when it was canceled.

Observing several dozen MI schools (mostly elementary schools in the United States) and conducting my own collaborative studies over 15 years, I have developed my own ideas about suitable MI applications. They are presented in the newly published *Intelligence Reframed: Multiple Intelligences in the 21st Century* and other publications, notably the three-volume *Project Zero Frameworks for Early Childhood Education* (Teachers College Press, 1998).

I have concluded that MI theory is most useful under two conditions. First, one must define a specific adult end state that is desired. If one wants adults who are civil to one another, one must develop the personal intelligences. If one desires adults who are sensitive to the arts, then one must develop one or more of the intelligences that are crucial to artistic performance and perception.

Second, one must decide which kinds of understanding are most important. I personally seek students who can understand the world in terms of the major disciplines—science, mathematics, history, the arts. This understanding is most likely to emerge from deep immersion in specific topics and, if one is willing to spend time on those topics, an MI approach becomes possible. In *The Disciplined Mind*, I examined three rich and complex topics: the theory of evolution, the music of Mozart and the Holocaust. I showed how understanding arises if one broaches these topics in a number of ways; draws on comparisons and analogies from complementary domains; and captures the key ideas in different symbol systems (e.g., linguistic, numerical, graphic, dynamic). Such applications show the power of an "MI approach" to serious scholarly work. Because students differ cognitively from one another, one reaches more students; and because disciplinary competence involves the capacity to represent the world in multiple ways, students gain a feeling for what it is like to be an expert.

What Is the Potential for the Future?

The idea of multiple intelligences is deceptively simple: "Human beings have eight or nine separate intelligences, not just the one or two that are highlighted in schools." Yet, even 20 years after proposing the theory, I discover new implications.

Part of my own interest centers on the theory itself. In recent years, I have pursued theoretical distinctions—for example, the differences among an intelligence (biological construct), a domain or discipline (an epistemological construct) and a field (a sociological construct). I have considered additional intelligences (naturalist, spiritual, existential). I keep up with data from biological, cognitive and anthropological sciences, monitoring with special interest studies using PET scans, functional MRIs and other ways of examining the human brain in vivo. With the theory as background, I have begun to investigate extraordinary humans (see *Creating Minds, Leading Minds, Extraordinary Minds* and most recently my ongoing collaborative study of persons who carry out "good work"). I also have been stimulated—if not always delighted—by critiques of the theory.

The world of educational practice has a life of its own. Practitioners often do not follow the theoretical work carefully; many individuals involved in MI have read neither my work nor that of other reliable writers. This state of affairs seems inevitable.

Clearly, ideas about multiple intelligences will continue to be put to use in a range of spheres (schools, other educational institutions, businesses) in many regions and countries. Nearly all applications will initially be superficial. I am interested in whether or not the experiments last; whether or not insight deepens over time; and, most of all, whether or not the MI ideas help to bring about greater success in the educational missions of the institutions that embrace them. Thus, I welcome institutions that are reflective about their practice and that actually employ measures ("hard" or "soft") indicating whether MI practices have been productive.

It is difficult to prove that improvements (or declines) in educational achievement are due to MI practices. In the real world of education, one cannot carry out controlled experiments where everything except MI practices is held constant. I have been criticized for not making more decisive statements about the efficacy of MI approaches. I have held off not because I doubt their efficacy, but because I know how difficult it is to prove that successes are due to MI. Surveys like SUMIT are pivotal.

I suspect that, as time passes, ideas about multiple intelligences will gain more acceptance among psychologists, though perhaps not my particular formulation. Psychologists favor research that relies on tests, while I question if this is the way to study the several intelligences.

Within formal education, I expect that a small number of schools will be developed based explicitly on MI beliefs and practices. More broadly, the work of my colleagues will encourage practitioners to enhance their teaching approaches, curricula and modes of assessment. Whether or not MI theory is explicitly acknowledged here is not important.

For me, the scientific and practical importance of MI theory centers on the idea that individuals differ cognitively from one another on a range of relatively autonomous dimensions. This variety is important in terms of evolution; it also spices up life for all of us on this planet. For most of recorded history, educators have ignored this variety in favor of uniform schools, where all students have been taught and assessed in the same way.

I may have helped to undermine this institutional reflex, but computers will do far more. Already it is possible to individualize education extensively by using software that engages a range of intelligences. In the future, that range will surely be extended; moreover, "intelligent systems" will adapt teaching to the strengths and interests of each student. In 50 years, our successors will laugh at the notion that there is but a single way to teach and assess. Instead, they will seek the best way to teach this concept or subject to this student and the best way for this student to demonstrate understanding. If MI theory turns out to have kept apace scientifically, it will be able to explain in biological, psychological and cultural terms why this educational approach has worked.

Traub's Opening Statement

What Is the Evidence for the Theory of Multiple Intelligences?

"Intelligence" is one of the most tendentious words in the English language. Human beings possess innumerable gifts and attributes, and yet we apply the word "intelligence" only to one tiny cluster of them. Howard Gardner has pointed out that the definition of this core of intellectual attributes changes from one era to the next, and is different in traditional cultures from what it is in technocratic ones. It seems to me that the simplest way to dispose of the unending debate about the validity of IQ tests and the like would be to concede that "intelligence" is not an entity but a construct, and then to proceed directly to an open discussion about which intellectual attributes we consider essential at this particular moment in history.

That, however, is not Gardner's position. Instead, he has argued, since the publication of *Frames of Mind* in 1983, that the pantheon of intelligence has been far too restrictive. To the traits measured by IQ tests, which he describes as "linguistic" and "logical-mathematical" intelligence, Gardner would add musical, spatial, interpersonal, intrapersonal, "bodily-kinesthetic" and what he calls "the naturalists' intelligence." Gardner says that he has considered, but not yet granted admission to, the "existential intelligence."

Unlike the proponents of "moral intelligence" or "emotional intelligence," Gardner has not used this terribly loaded term, intelligence, simply for effect; he has furnished a set of objective criteria that may be used to distinguish it from a mere aptitude. In *Frames of Mind*, Gardner writes that he calls a trait an "intelligence" only if "it can be found in relative isolation in special populations," "may become highly developed in specific individuals or in specific cultures," and if "experts in particular disciplines can posit core abilities that, in effect, define the intelligence." Elsewhere, he distinguishes between information-processing skills and values, or attributes of character.

Gardner's signal achievement is to bring the findings of brain research into a world hitherto defined by test outcomes. He premises his "multiple intelligence" theory on the "modular" or "vertical" picture of the brain that is now widely accepted in neuroscience. We now know that mental activities are parceled out into separate regions of the brain. This, Gardner says, argues for a view of intelligence as consisting of relatively autonomous faculties, rather than for the traditional view of a bundle of aptitudes highly correlated with one another and together constituting a larger whole called "general intelligence," or G.

Most psychologists who study intelligence scoff at Gardner's work, pointing out that he has merely posited his various intelligences; neither he nor anyone else has yet done the psychometric work that would be needed to prove that they are distinct, rather than aspects of one another or of something else. Gardner's argument for the relative autonomy of intellectual gifts is broadly accepted, up to a point. Brain researchers recognize, for example, that many types of memory and even learning capacity are scattered throughout the brain. Research in brain-damaged patients has shown that it is possible to lose one core intellectual ability while others remain unaffected. But to abandon G is to accept a view of the brain as having little or no executive capacity to direct and integrate the mind's activity. Gardner has suggested, alternatively, that each intelligence might, in effect, have its own G, but many cognitive psychologists balk at this idea. Can we really say that "musical intelligence" is so autonomous from, and uncorrelated with, the skills that constitute G?

Gardner's "multiple intelligence" theory, perhaps like all theories about intelligence, begins in science and ends in cultural politics. G is a construct used to explain why the various aptitudes measured by IQ tests correlate with one another so highly. It would hardly be a surprise if aptitudes not measured by IQ tests—say, the bodily-kinesthetic ones—are completely independent of IQ scores. We measure some qualities and not others—and call the ones we measure "intelligence"—because we think they matter more in determining success in a technocratic culture, as they have since Binet's time. They do not, of course, determine human value, though we, and the psychometricians, often talk as if they do.

Gardner, who once trained to be a concert pianist, recoils at this narrow conception of man-the-logician. In one of his first books, *The Arts and Human Development*, he rebuked the developmental psychologist Jean Piaget for studying only "those mental processes that culminate in scientific thought." Creative thought, he pointed out, is just as fundamental a constituent of human behavior as logic. Gardner is, at bottom, a moral philosopher; he wants to change the way we measure human worth. That is a profound project, if perhaps only incidentally a scientific one.

What Has Been the Impact on Education?

Frames of Mind landed in the school world with the force of revelation. Though Gardner had written scarcely a word about pedagogy, "MI schools" began to spring up within a few years of the book's publication, and were soon followed by MI teaching guides, consultants and publishers. Gardner had offered a theoretical framework and a scientific explanation for an intuition shared by many teachers: that children have different ways of learning and must be approached with different kinds of teaching. Gardner also had validated progressive education's focus on the learner rather than on the knowledge being imparted, a principle that had already attained the status of orthodoxy in education schools and among many teachers.

The common thread of MI schools is the use of what Gardner calls "multiple entryways" to curricular knowledge. Thus, at one elementary school described in an educational journal, students learning about photosynthesis "might act out the process at one [learning] station, read about it at another station, and, at others, sing about photosynthesis, chart its processes and, finally, reflect on events that have transformed their lives, just as chloroplasts transform the life cycle of plants." Two things are at least allegedly happening here: The school is stimulating all of the child's intelligences, and the child is mobilizing those various faculties to gain a deeper understanding of subject matter.

Gardner has made a point of writing that everything ought not be taught seven or eight different ways, and many schools find ways of integrating the MI philosophy without being so literal-minded about it. Nevertheless, the question this and similar descriptions raises is: Have many students been failing to learn about photosynthesis, or fractions, or American history, because they happen not to have the mix of intelligences suited to the traditional text-oriented curriculum? If this were true, one would have to imagine that the Asian and European students who regularly outperform Americans were benefiting from a more personalized form of pedagogy—which, of course, they are not. The educational system in both Taiwan and France, for example, is determinedly knowledge-centered, and it is unambiguous about the primacy of logic and language.

But perhaps this is the wrong question. Progressive reformers argue that our schools aim too low, that the knowledge imparted even by apparently successful schools is superficial; they draw a sharp distinction between "knowledge" and "understanding." In *The Unschooled Mind*, Gardner observes that even students at the highly selective Massachusetts Institute of Technology provide childishly naive explanations of ordinary real-world phenomena. What they have learned with paper and pencil—and their logical intelligence—has left them with a merely formulaic grasp of the subject matter. The mark of true understanding, Gardner writes, is the ability to make "multiple representations" of a given subject. He would have us teach subjects in their own media as well as abstractly.

This provocative thought about true understanding echoes the work of educational philosophers such as Seymour Papert. Nevertheless, American schools are in crisis not because even the most expert students suffer from a flimsy grasp of subject matter, but because so many students have almost no grasp of subject matter, nor do they have the skills they need to remedy the situation. Indeed, the progressive distinction between knowledge and understanding seems peculiar and almost perverse. A persuasive case can be made that American students are doing poorly as a result of the progressive aversion to a specified curriculum, not because of a Gradgrindish preoccupation with "drill and practice." Gardner's harshest critics, in fact, are not the psychometricians, but educational traditionalists such as E. D. Hirsch and educational psychologists such as Harold Stevenson, who believe that Gardner has things backwards.

The proliferation of the MI model may well have the positive effect of giving music and art—and even physical education—a more central place in the curriculum. It may, on the other hand, have the harmful effect of reducing the sense of urgency needed to ensure that all children master basic skills. MI theory can offer a pretext to put self-esteem ahead of the hard work of learning: Rather than harp on what a child doesn't do well—such as reading or math—why not focus on his or her own special gifts? Isn't singing about photosynthesis, or putting on a play about it, a valid form of learning? Many schools will not need much encouragement in this regard. The guiding principle of one well-known MI school is, "Who you are is more important than what you know." That seems like an excellent formula to ensure that you do not know much.

What Is the Potential for the Future?

Gardner believes strongly that we are living at the tail end of an exhausted paradigm of human nature. Invoking the work of Binet, he asks why we should continue to live with a definition of intelligence based on a "scholastic skill—what it meant to be a good bureaucrat a hundred years ago." Why, he asks, do we continue to distribute so precious a commodity as a seat in our greatest universities on the basis of minute distinctions in performance on the SAT exams, our ubiquitous surrogate for the IQ test? Gardner believes that the walls of the IQ meritocracy are crumbling (though the psychometric establishment has not yet noticed). "There is a new definition of human beings, cognitively speaking," he has grandly said. "Socrates defined man as a rational animal; Freud defined him as an irrational animal. What MI theory says is that we are the animal that exhibits the eight and a half intelligences." ("Existential" intelligence is the half.)

This raises two basic questions: Is the old paradigm really on its last legs, and what would it mean to have a new one based on Gardner's conception?

Of course, the reason why we still live by Binet's definition of intelligence is that yesterday's good bureaucrat is today's good midlevel executive or corporate lawyer. This is what former Secretary of Labor Robert Reich means when he says that the skill increasingly in demand in the world is that of "symbolic analyst." An increasing portion of the population works with abstractions for a living; that is why the minimal definition of school success has advanced so far beyond the achievement of literacy. It is quite true that the universalization of the computer will automate much of the brute work of logic; but the proliferation of spreadsheets will probably have the effect of making spreadsheet-type thinking more conventional, not less so.

At the same time, we no longer view rationality as the sine qua non of modernity. That era seems to have lasted about from the time of Sir Francis Galton (1822–1911) to the time of Vietnam-era Secretary of Defense Robert McNamara. We have been forcibly disabused of the wisdom of such wise men, and of their machines. So Gardner is quite right to say that the sense of human personality has changed. The immense popularity of Daniel Goleman's theory of "emotional intelligence" shows how widespread is the wish for a less hyperrational view of character. The same may be said for the incredible reception that Gardner's work has enjoyed far from the worlds of science. The restless and often rootless search for meaningful forms of faith reflects the inadequacy of the old secular gods. We grow more estranged from rationality as ever more rationality is demanded of us.

Will we turn instead to a more diverse, less hierarchical view of human gifts? The answer, to a certain extent, must be "no." So long as the commanding heights of the culture require gifted symbolic analysts, the institutions that serve as sorting devices for the elite—say, Harvard University—will keep selecting candidates based mostly on Binet's hundred-year-old criteria. In certain quarters, Gardner's paradigm will continue to be seen as a pretext for refusing to accept distinctions of merit. And yet it is not implausible that our sense of merit will change—that the grade point average of that girl applying to Harvard will be measuring her reflectiveness, her social gifts, her classificatory skills, her aesthetic sense, as much as her linguistic and mathematical talents. We may, in effect, come to think of well-roundedness as itself the supreme expression of merit.

One is tempted to growl, as Robert Frost once allegedly did, "Why d'ya wanna be well-rounded—you wanna roll downhill?" But perhaps one ought not be so curmudgeonly in defense of the hegemony of logic and language. We are, after all, a Philistine culture; we would be less so in Gardner's ideal world. We would also, presumably, be more self-knowing, more socially adroit, more aware of our bodies. But we would also, necessarily, be less something—a little less intellectually nimble, perhaps. Gardner likes to poke fun at what he calls the Alan Dershowitz model of intelligence. In Gardner's future, we will have more well-rounded folks and fewer like Harvard law professor Alan Dershowitz. One need not excessively respect Claus von Bulow's lawyer to say that this does not sound like the answer to America's prayers.

Gardner's Rejoinder

In his provocative statement, James Traub wears a number of hats (cultural critic, science reporter, capsule biographer) and assumes a number of tones (respectful, neutral, ironic). Rather than responding to each, I have chosen to clarify some points for both Traub and our readers.

Science and Politics

Traub suggests that all theories of intelligence begin in science and end in cultural politics. It may be true that, in our culture, theories of intelligence are drawn upon to undergird social recommendations (e.g., *The Bell Curve*). But one must draw a sharp distinction between those who encourage blurring of boundaries and those who strive to keep the realms discrete. For example, Arthur Jensen and Hans Eysenck are both psychologists who believe in the explanatory power of G. Throughout his career, Eysenck moved almost too effortlessly between psychometrics and policy; but after one disastrous foray, Jensen has focused on G, avoiding policy debates altogether.

As an educational reformer and a citizen, I have described the schools I favor and the society I cherish. But I have sought to keep my scholarly analysis of intelligence separate from my policy recommendations. Indeed, I have insisted that policy never follows directly from scientific discourse. And I have always emphasized that the various intelligences are amoral in themselves: One can use interpersonal intelligence to resolve a dispute or to manipulate groups.

Intelligences and Domains

Many writers, including me, have sometimes confused "intelligences" with "domains" or "disciplines." An intelligence is a biopsychological potential that can be drawn on for a variety of skills or roles. A domain or discipline is a culturally recognized area of performance whose practitioners can be arrayed in terms of expertise. Thus spatial intelligence can be exploited in domains like chess playing, sailing and surgery; and the domain of law draws, variously, on linguistic, logical-mathematical and interpersonal intelligences. Even the terms "musician" and "musical intelligence" are not interchangeable; musicians exploit several intelligences in composing or performing.

This conceptual distinction is important. Ultimately, as educators or citizens, we should not care which intelligences individuals are using; the important goal is to achieve reasonable performances in domains that matter. An individual with modest logical-mathematical intelligence can still learn to carry out mathematical operations; he will have to rely more than others do on linguistic, spatial and perhaps bodily intelligences. The challenge for educators, and for the individual, is to marshal and choreograph intelligences to obtain proficiency. Contrary to Traub's implications, there is no reason whatever to avoid teaching literacy or the disciplines to youngsters deficient in a given intelligence. MI theory simply provides a convenient way of analyzing how best to approach a domain when customary teaching practices fail.

Indeed, unless we literally peer inside the mind/brain, we have no way of knowing which intelligences an individual is actually using. You are reading these words (a linguistic skill), but you may be representing them mentally using various intelligences (ranging from an interpersonal debate between the journalist and the academic to a numerical scorecard to a spatial layout of arguments). An MI classroom in the United States may publicly evoke different intelligences in teaching about photosynthesis; but for all we know, students in France or Taiwan may also be encoding such biological systems in a spatial or naturalistic way. Nor should we assume that other societies ignore individual differences. Early education in Japan focuses on social and personal development, and many Japanese families use individual or group tutoring to supplement schooling.

Symbol Analysts and Anti-rationality

In various writings, Traub has suggested that I am opposed to logic and rationality. This is simply untrue. I try to operate according to those canons and hope others will as well. While I poke gentle fun at my colleague Alan Dershowitz, I have never denigrated his powerful mind; I have only pointed out that many schools and colleges are currently set up to select one "frame of mind" above others. Fortunately, recognizing that standardized tests predict but a modest proportion of future success, some selective institutions welcome other samples of student work.

Throughout much of this century, it is true, individuals who are expert in "symbol analysis" have been at a premium in our society. It is not unreasonable to assume that they are using linguistic and/or logical intelligence, but, again, we do not know this for sure. Charles Schwab, for example, takes a back seat to few in his financial acuity, yet he speaks freely about his great difficulty in learning to read. There are many ways to conceive of, and analyze, a financial market.

Two other points. First of all, it takes various gifts and mixes of talents to have a productive society. As Traub points out, the canonical "best and the brightest" have sometimes inspired disastrous policies. Second, the kinds of abilities at a premium in a society can change, sometimes quite swiftly. Inventions like the printing press, the computer and the cinema bring certain intelligences to the fore while (at least temporarily) de-emphasizing others. The smarter machines become, in the Binet sense, the more emphasis society is likely to place on intelligences that transcend standard computation. And so, just as evolution is friendly to diversity, we are well advised not to put too much stock in the nurturance of one or two intelligences.

Knowledge (Facts) and Understanding

Traub imposes on me a dichotomy that I do not make and will not accept. Of course, it is not possible to understand a topic unless you have considerable knowledge, including factual. Who could maintain otherwise? The distinction that I make, most recently in *The Disciplined Mind*, is twofold. First and foremost, I believe that facts ought to be picked up through intensive, deep study of consequential issues. Not only will one have the motivation to learn the facts; one will also be in a position to put them together in meaningful ways, to recall them and to master vital disciplinary ways of thinking. Absent such organization and mission, the facts remain "inert knowledge"—available, perhaps, for the next test but likely to be forgotten and, in any case, not leading to disciplined thought. Second, I have little sympathy for the delineation of a specific canon—specific facts, theories or books that all must master. The most important disciplinary habits of mind can be obtained from a variety of topics; and once they have been obtained, one is free to master any canon or create one's own.

False Dichotomies

While I have broad sympathy with the progressive tradition in education (of philosopher John Dewey), I would be happy to chuck the distinction between Progressive and Traditionalist. It's too often used to vilify, rather than to illuminate. As for the contrast between "who you are" and "what you know," after the Littleton, Colorado, school massacre and similar tragedies, I hope that we do not consider this an either/or choice.

And Finally ...

1. Discussion of an "executive function" is a separate issue from the existence or provenance of G.
2. Creativity should not be contrasted with science or logic. There are creative scientists and logicians, and (all too many) noncreative artists.
3. Never have I ever stated or implied that basic skills (the three Rs) should not come first. I continuously insist that they must.
4. As a youngster, I was a serious pianist but never trained for the concert stage. I still, however, have dreams.

Traub's Rejoinder

Is it really true that "If a person is strong in telling stories [or] solving mathematical problems ... one does not know if comparable strengths (or weaknesses) will be found in other areas"? Does this assertion's contradiction of our intuitive sense of the world only show how captive we are to the psychometric tradition? In *Frames of Mind*, Gardner pulls together evidence from an extraordinarily wide variety of fields to show that these faculties have distinctive evolutionary histories, patterns of pathology and brain localization. But one is left wondering exactly how autonomous is "relatively autonomous."

We know, of course, of people who are especially gifted in "understanding the motivations of others" or "tracing the transformations of a fugal theme." But do we know of people who have such talents and yet lack the "general intelligence" measured by IQ tests? I do not. Relative autonomy does not, after all, preclude correlation. As Stephen Ceci, a developmental psychologist at Cornell University, says, "If you tested people in track and field and you found someone who was really outstanding in one particular event, like the hurdle or the high jump, you'd also find that they were above average in all the others."

Learning for Living

Let us, however, stipulate for argument's sake that Gardner is right at the level of neuroscience and psychometrics. What happens when we apply his thinking to the world around us? What does it mean to think of logic and language as merely two of eight distinct and epistemologically equal gifts? Gardner has observed that in some traditional cultures—in western Africa, for example—musical intelligence is counted the *summum bonum* of human achievement. In the future, he observes, with computers doing the work of logic for us, we ourselves may put more store by social graces—that is, by "interpersonal" intelligence. Gardner has demonstrated that cultures adapt the concept of intelligence to their own purposes. But what about our present purposes? Since we do not get around in dugout canoes, learning to navigate by the stars is not terribly relevant to most of us. It may be a real gift, and it may even rest on a distinct intelligence, but, so far as our acculturating institutions go, it is pretty marginal.

Americans have a terrible problem with their principal acculturating institution—the public schools. According to last year's National Assessment of Educational Progress, 38 percent of fourth-graders read at a "below basic" level—the same percentage as score "at or above proficient." We perform middlingly in the lower grades, and at or near the bottom in the upper grades, on all of the indices of the Third International Mathematics and Science Study.

Harold Stevenson, a psychologist at the University of Michigan, compared the performance of students at demographically similar schools in the United States, Japan and Taiwan and found that, among fifth-graders, only 4 percent of Chinese students and 10 percent of Japanese students did as poorly on math as the average American; the figures were almost as bad in 11th grade. Stevenson explained the huge discrepancy by noting that the American students spent far less time on academic work than did the others; that American teachers, lacking a national curriculum or standards, rarely worked together; that American parents were far more easily satisfied by their children's work than were Asian parents; and that while Asians stress hard work, Americans believe that success is due to innate ability. Gardner's multiple intelligence theory, of course, offers a new and progressive-sounding version of the American preoccupation with innate ability.

A Stand on Standards

No one can question Gardner's own intellectual seriousness, but his pedagogy seems designed to address some problem other than the one that our schools face. He has criticized standardized tests generally, and recently he wrote skeptically of New York State's intellectually ambitious fourth-grade reading and writing test, asserting that we must not lose sight of "the most crucial skills: love of learning, respect for peers and good citizenship." Like many progressives, he opposes a national curriculum or national standards. His criticism of the hegemony of logic and language seems oddly malapropos in today's schools. Stevenson observes:

> If you're giving this child excessive feelings of accomplishment because the child is good at music, and not giving the child the sense of need to become accomplished in abstract intelligence, then you're depriving that intelligence for important parts of his or her future development.

Stevenson and other traditionally minded reformers, such as Diane Ravitch, argue that Gardner has given aid and comfort to educators who want to excuse mediocre performance and justify low standards. But Gardner's actual influence on the schools is certainly more complicated than that. Many ambitious and thoughtful teachers and administrators have been inspired by his books and have seen them as a justification for holding children to the highest, not the lowest, standards.

The two "MI" schools I have visited had their eccentricities, but they expected a high level of academic performance from children.

At Governor Bent Elementary School in Albuquerque, fifth-graders who had been defined as gifted, and yet were deficient in reading and writing—itself a kind of multiple intelligence diagnosis—had been assigned sophisticated architecture projects, including the construction of an ant colony that could withstand flood, hurricane and enemy attack, and the building of a robot with a part that could rotate 360 degrees. "If I was offering the kids another kind of curriculum," said their teacher, Marleyne Chula, "you would not see them engaged like this." But children at Governor Bent are also expected to bark out the names of painters, composers and U.S. presidents on command—a most un-Gardnerian faith in "decontextualized facts." Perhaps the school succeeds out of sheer inconsistency.

In *The Disciplined Mind*, Gardner made the startling admission that he could imagine approving of a national curriculum—save for his fear that a "Jesse Helms" would take control of it. It might be a salutary exercise for Gardner himself to write such a curriculum—not an airy "pathway of understanding," as he limns in *The Disciplined Mind*, but a year-by-year, subject-by-subject syllabus. Some unexpected people might be happy to embrace it.

Gardner's Closing Statement

Like the imminent hanging described by Dr. Johnson, a 500-word limit concentrates the mind. I am now convinced that Traub is not principally concerned with multiple intelligences, nor with an ideal education. Unrelentingly pragmatic, he believes American schools are not good and need to improve. He sees MI as at best a benign grace note, at worst destructive.

I am no apologist for American education. I share Traub's impatience with its often mediocre accomplishments (though I may have more empathy for the difficulty faced by many teachers). But I am convinced that there is no one right way to achieve better schools. Widely divergent schools succeed in different countries and in ours. Our much-admired colleges are distinguished by their diversity. In *The Disciplined Mind*, I concluded that so varied a country as ours cannot agree on a single gritty and challenging curriculum. We will either fight endlessly (Is creationism a theory?) or settle for pabulum. Accordingly, I sketched six alternative K–12 pathways. I even conceded that I would rather send my children along a pathway whose philosophy is not appealing but that pursues its philosophy consistently than a pathway that applies "my" philosophy poorly.

A Personal Vision

The Disciplined Mind outlines my personal vision: a pathway that begins by teaching the three Rs; assigns project work that addresses essential questions and engenders motivation to pursue those questions; requires work that begins to introduce the major disciplinary divisions; and then involves a deep exploration of several major disciplines: science, history, mathematics, the arts and considerations of ethics. Students will acquire facts, but not acontextually; rather, they will learn facts naturally as they investigate consequential matters and are stimulated to read, write and explore widely. If they lack a fact, they can look it up or use tomorrow's Palm Pilot that issues facts on oral request.

For me, the irreplaceable essence of an education is a mastery of the major ways of thinking that have developed over the centuries in several disciplines. These provide the "mental furniture" that alone can make sense of past materials and allow one to understand new materials, phenomena, events. Absent the disciplinary capacity to apply or assimilate information, one has only "inert knowledge"—Christmas tree ornaments without the fundamental tree and branches.

In *The Disciplined Mind*, I detailed an educational investigation of three major topics—evolution, the music of Mozart and the Holocaust. In the process, one would learn what it is like to understand consequential materials and would also be introduced, respectively, to the disciplines of biological science, music and history. I elaborated on how the fact of our multiple intelligences can be used to make this material accessible for students and ultimately yield a rich set of representations of a topic. My discussion provides a curricular model for these topics. I hope that it will inspire others to create and to "own" their own curricula. But I am not a curriculum writer, any more than James Traub is a research social scientist.

Notes as the noose tightens:

1. Traub seems fixated on how to improve our standing in international comparisons. I am concerned with the formation of disciplined minds, which, I believe, are the best preparation for the future. Current tests (and ways of testing) are inadequate.
2. Brain study will help us understand the mind, but you can never go directly from brain to education. Education involves considerations of epistemology (factual accretion versus disciplinary analysis) and values (wide cultural literacy versus in-depth understanding of key topics). These decisions must be made by communities, not by scientists or pundits.

Traub's Closing Statement

I have to wonder why I would bother to pick a fight with someone as thoughtful, and as genuinely creative, as Howard Gardner. (Here I am in my "respectful" mode.) I recognize that this probably has less to do with his own work than with the effect I fear it will have on the world. It is absolutely true that Gardner has resisted the idea that intelligence can be something other than a form of cognitive processing—that it can be inherently moral, for example. It is also true that he has stressed his own faith in intellectual rigor. But many of his readers have no such faith and are not so inclined to distinguish between the cognitive and the noncognitive. And so I cannot feel much confidence about how they will use the concepts of "bodily-kinesthetic intelligence" or "the personal intelligences." I fear they will be used at least as excuses, and not just as opportunities. And though I do not wish to sound hard-hearted, I do not agree that the lesson of Littleton was that schools are failing to teach that who you are is more important than, or as important as, what you know. I do, however, believe that one very important lesson of Littleton was that if you want to have a culture of respect and mutual understanding, you should not have 1,800 kids in a high school.

What Should We Measure?

I cannot help bridling at the phrase, "An individual with modest logical-mathematical intelligence … ." Of course, I do not dispute the idea that intelligence is significantly heritable. But is there not something retrograde about going from the idea of IQ to the idea of seven or eight IQs? Would we not be better off focusing on effort, rather than innate ability? I agree with Gardner that the SATs have vastly too much power in our society, but that is because they measure, or pretend to measure, underlying capacities rather than disciplinary knowledge. We should encourage and reward effort by measuring what students have actually learned. This, by the way, is another argument for a national curriculum, or at least for national "standards."

I honestly do not see why the particulars of a specified curriculum would inevitably be reduced to "inert knowledge." That has to do with how one teaches, not what one teaches. At a school I visited in Cambridge that follows E. D. Hirsch's highly detailed Core Knowledge curriculum, kindergartners were learning about Mussorgsky while dancing around to "Pictures at an Exhibition" (very MI), third-graders were reading books to one another and sixth-graders were attending a local play in order to bring the practice of commedia dell'arte to life. Is it really unreasonable to think that the lion of factualness can lie down with the lamb of understanding?

5

The Teen Brain

Primed to Learn, Primed to Take Risks

By Jay N. Giedd, M.D.

Jay N. Giedd, M.D., is a practicing child and adolescent psychiatrist and chief of brain imaging at the National Institute of Mental Health (NIMH) Child Psychiatry Branch. In an ongoing longitudinal study of more than 2,000 people, Dr. Giedd combines neuroimaging, genetics and psychological testing to explore the developing brain in health and in illness.

During adolescence the brain's ability to change is especially pronounced—and that can be a double-edged sword. Jay N. Giedd, a child and adolescent psychiatrist who specializes in brain imaging at the National Institute of Mental Health, points out that the brain's plasticity allows adolescents to learn and adapt, which paves the way for independence. But it also poses dangers: Different rates of development can lead to poor decision-making, risk-taking and, in some cases, diagnosable disorders.

ACROSS CULTURES AND MILLENNIA, the teen years have been noted as a time of dramatic changes in body and behavior. During their teens, most people successfully navigate the transition from dependence upon family to self-sufficiency as an adult member of the society. However, adolescence is also a time of increased conflicts with parents, mood volatility, risky behavior and, for some, the emergence of psychopathology.

The physical changes associated with puberty are conspicuous and well described. The brain's transformation is every bit as dramatic but, to the unaided eye, is visible only when it causes new and different behavior. The teen brain is neither broken nor defective. Rather, it is wonderfully optimized to promote our success as a species.

Beginning in childhood and continuing through adolescence, dynamic processes drive brain development, creating the flexibility that allows the brain to refine itself—to specialize and sharpen its functions for the specific demands of its environment. Maturing connections pave the way for increased communication among brain regions, thus enabling greater integration and complexity of thought. When what we call adolescence arrives, a changing balance between brain systems involved in emotion and regulating emotion spawns increased novelty seeking and risk-taking and causes a shift toward peer-based interactions.

These behaviors, found in all social mammals, encourage us to separate from the comfort and safety of our families in order to explore new environments and to seek unrelated mates.[1] These potentially adaptive behaviors also pose substantial dangers, however, especially when mixed with modern temptations and easy access to potent substances of abuse, firearms and high-speed motor vehicles.

In many ways adolescence is the healthiest time of life. The immune system, resistance to cancer, tolerance for heat and cold and several other variables are at their peak. Despite physical strengths, however, illness and mortality increase by 200 to 300 percent. As of 2005, the most recent year for which statistics are available, motor vehicle accidents, the No. 1 cause of death among adolescents in the United States, accounted for about half of deaths. Nos. 2 and 3 were homicide and suicide.[2] Understanding this healthy-body, risk-taking-brain paradox will require greater insight into how the brain changes during this period of life. Such enhanced understanding may help to guide interventions when illnesses emerge or to inform parenting or educational approaches to encourage healthy development.

Adolescent Neurobiology: Three Themes

The brain, the most protected organ of the body, has been particularly opaque to investigation of what occurs during adolescence. But now the picture emerging from the science of adolescent neurobiology highlights both the brain's capacity to handle increasing cognitive complexity and an enormous potential for plasticity—the brain's ongoing ability to change. The advent of structural and functional magnetic resonance imaging (MRI), which combines a powerful magnet, radio waves and sophisticated computer technology to provide exquisitely accurate pictures of brain anatomy and physiology, has opened an unprecedented window into the biology of the brain, including how its tissues function and how particular mental or physical activities change blood flow. Because MRI does not involve ionizing radiation, it is well suited for pediatric studies and has launched a new era of neuroscience. Three themes emerge from neuroimaging research in adolescents:

1. Brain cells, their connections and receptors for chemical messengers called neurotransmitters peak during childhood and decline in adolescence.

2. Connectivity among brain regions increases.

3. The balance among frontal (executive-control) and limbic (emotional) systems changes.

These themes appear again and again in our studies of the biological underpinnings of cognitive and behavioral changes in teenagers.

Theme 1: Childhood Peaks Followed by Adolescent Declines in Cells, Connections and Receptors

The brain's 100 billion neurons and quadrillion synapses create a multitude of potential connection patterns. As teens interact with the unique challenges of their environment, these connections form and re-form, giving rise to specific behaviors—with positive or negative outcomes. This plasticity is the essence of adolescent neurobiology. It underlies both the enormous learning potential and the vulnerability of the teen years.

Neuroimaging reveals that gray matter volumes—which reflect the size and number of branches of brain cells—increase during childhood, peak at different times depending on the location in the brain, decline through adolescence, level off during adulthood and then decline somewhat further in senescence. This pattern of childhood peaks followed by adolescent declines occurs not only in gray matter volumes, but also in the number of synapses and the densities of neurotransmitter receptors.[3] This one-two punch—overproduction followed by competitive elimination—drives complexity not only in brain development, but also across myriad natural systems.

Theme 2: Increased Connectivity

Many cognitive advances during adolescence stem from faster communication in brain circuitry and increased integration of brain activity. To use a language metaphor, brain maturation is not so much a matter of adding new letters as it is one of combining existing letters into words, words into sentences and sentences into paragraphs.

The term *connectivity* characterizes several neuroscience concepts. In anatomic studies connectivity can refer to a physical link between areas of the brain that share common developmental trajectories. In studies of brain function, connectivity describes the relationship between different parts of the brain that activate together during a task. In genetic studies it refers to different regions that are influenced by the same genetic or environmental factors. All of these types of connectivity increase during adolescence.

In structural MRI studies of brain anatomy, connectivity, as indicated by the volume of white matter—bundles of nerve cells' axons, which link various areas of the brain—increases throughout childhood and adolescence and continues to grow until women reach their 40s and men their 30s. The foundation of this increase in wiring is myelination, the formation of a fatty sheath of electrical insulation around axons. This sheath speeds conduction of nerve impulses. The increase is not subtle—myelinated axons transmit impulses up to 100 times faster than unmyelinated axons. Myelination also accelerates the brain's information processing via a decrease in the recovery time between firings. That allows up to a 30-fold increase in the frequency with which a given neuron can transmit information. This combination—the increase in speed and the decrease in recovery time—is roughly equivalent to a 3,000-fold increase in computer bandwidth.

However, recent investigations into white matter are revealing that myelin has a much more nuanced role than a simple "pedal to the metal" increase in transmission speed. Neurons integrate information from other neurons by combining excitatory and inhibitory input. If excitatory input exceeds a certain threshold, the receiving neuron fires and initiates a series of molecular changes that strengthen the synapses, or connections, from the input neurons. This process, which forms the basis of learning, has been summarized, "Neurons that fire together wire together." In order for input from nearby and more distant neurons to arrive simultaneously, the transmission must be exquisitely timed. Myelin is intimately involved in the fine-tuning of this timing, which encodes the basis for thought, consciousness and meaning in the brain.

The dynamic activity of myelination during adolescence reflects how much new wiring is occurring.

On the flip side, recent research reveals that myelination also helps close the windows of plasticity by inhibiting axon sprouting and the creation of new synapses.[4] Thus, as myelination proceeds, the circuitry that is used the most becomes faster, but at the cost of decreased plasticity.

Advances in imaging techniques, such as diffusion tensor imaging (DTI) and magnetization transfer (MT) imaging, have helped spark interest in these processes because they allow researchers to characterize the directions of axons and the microstructure of white matter. These new techniques further confirm an increase in white matter organization during adolescence, and in specific brain regions, this increase correlates with improvements in language,[5] reading,[6] ability to inhibit a response[7] and memory.[5]

Functional magnetic resonance imaging (fMRI) studies also consistently demonstrate increasing and more efficient communication among brain regions during childhood and adolescence. We can measure this communication by comparing the activation of different brain regions during a task. In studies assessing memory[8] and resistance to peer pressure,[9] the efficiency of communication in the relevant circuitry was a better predictor of how teens performed than was a measurement of metabolic activity in the regions involved.

These lines of investigation, along with electroencephalography (EEG) studies indicating increased linking of electrical activity in different brain regions, converge to establish a fundamental maturation pattern in the brain: an increase in cognitive activity that relies on tying together and integrating information from multiple sources. These changes allow for greater complexity and depth of thought.

Theme 3: Changing Frontal/Limbic Balance

The relationship between earlier-maturing limbic system networks, which are the seat of emotion, and later-maturing frontal lobe networks, which help regulate emotion, is dynamic. In order to understand decision-making during adolescence, we must appreciate the interplay between limbic and cognitive systems. Psychological tests are usually conducted under conditions of "cold cognition"—hypothetical, low-emotion situations. However, real-world decision-making often occurs under conditions of "hot cognition"—high arousal, with peer pressure and real consequences. Neuroimaging researchers continue to discern the different biological circuitry involved in hot and cold cognition and are beginning to map how the parts of the brain involved in decision-making mature.

Frontal lobe circuitry mediates executive functioning, a term encompassing a broad array of abilities, including attention, response inhibition, regulation of emotion, organization and long-range planning. Structural MRI studies of cortical thickness indicate that areas involved in high-level integration of input from disparate parts of the brain mature particularly late and do not reach adult levels until the mid-20s.[10]

Across a wide variety of tasks, fMRI studies consistently show an increasing proportion of frontal versus striatal or limbic activity as we progress from childhood to adulthood. For example, among 37 study participants ages 7 to 29, the response to rewards in the nucleus accumbens (related to pleasure-seeking) of adolescents was equivalent to that in adults, but activity in the adolescent orbitofrontal cortex (involved in motivation) was similar to that in children.[11] The changing balance between frontal and limbic systems helps us understand many of the cognitive and behavioral changes of adolescence.

Normal Changes versus Pathology

One of the greatest challenges for parents and others who work with teens is to distinguish exasperating behavior from genuine pathology. Against the backdrop of healthy adolescence, which includes a wide range of mood fluctuations and occasional poor judgment, is the reality that many types of pathology emerge during adolescence—anxiety disorders, bipolar disorder, depression, eating disorders, psychosis and substance abuse, among others. The relationship between normal neurobiological variations and the onset of psychopathology is complicated, but one underlying theme may be that "moving parts get broken." In other words, brain development may go awry, thus predisposing adolescents to disorders. Although neuroimaging techniques are helping researchers establish correlations between brain structure or function and behavior, scientists have not firmly established a link between typical behavioral variations and psychopathology. For example, the neural circuitry underlying teen moodiness may not be the same circuitry involved in depression or bipolar disorder. A greater understanding of the relationships between specific adolescent brain changes and their specific cognitive, behavioral and emotional consequences may provide insight into prevention or treatment.

In the meantime, late maturation of the prefrontal cortex, which is essential in judgment, decision-making and impulse control, has prominently entered discourse affecting the social, legislative, judicial, parenting and educational realms. Despite the temptation to trade the complexity and ambiguity of human behavior for the clarity and aesthetic appeal of colorful brain images, we must be careful not to over-interpret neuroimaging findings as they relate to public policy.

Questions about maturity and the age of consent are particularly enmeshed in political and social contexts. For example, in the United States today a person must be at least 15 to 17 years old (depending on the state) to drive; at least 18 to vote, to buy cigarettes or to be in the military; and at least 21 to drink alcohol. The minimum age for holding political office varies as well: Some municipalities allow mayors as young as 16, and the minimum age for governors ranges from 18 to 30. (On the national level, 25 is the minimum age to be a member of the U.S. House of Representatives; a senator or president must be at least 35.) The age of consent to sexual relations varies worldwide, from puberty (with no specific age attached) to 18. In most cultures the age at which a female may consent to sexual relations is lower than the age for a male. In the United States, the legal age to consent to sexual intercourse varies by state from 14 to 17 for females and from 15 to 18 for males. These demarcations reflect strong societal influences, and they do not pinpoint a biological "age of maturation." For instance, the age of majority (legal adulthood) was increased from 15 to 21 in 13th-century England because soldiers needed to be strong enough to bear the weight of protective armor and to acquire the necessary skills for combat. Societal influences also contributed to the 26th Amendment to the U.S. Constitution, which in 1971 lowered the voting age from 21 to 18 to address the discrepancy between being able to be drafted and being able to vote. Delineating the proper role of developmental neuroscience, particularly neuroimaging, in informing public policy on age-of-consent issues will require extensive deliberation with input from many disciplines.

From the perspective of evolutionary adaptation, it is not surprising that the brain is particularly changeable during adolescence—a time when we need to learn how to survive independently in a wide variety of environments. Humans can survive in the frozen tundra of the North Pole or in the balmy tropics on the equator. With the aid of technologies that began as ideas from our brains, we can even survive in outer space. Ten thousand years ago—a blink of an eye in evolutionary terms—our brains might have been optimized for hunting or for gathering berries. Now our brains might be fine-tuned for reading or programming computers. This incredible changeability, or plasticity, of the human brain is perhaps the most distinctive feature of our species. It makes adolescence a time of great risk and great opportunity.

6

Video Games Affect the Brain—for Better *and* Worse

By Douglas A. Gentile, Ph.D.

Douglas A. Gentile, Ph.D., directs the Media Research Lab in the psychology department at Iowa State University, where he conducts research on the media's effects on children and adults. He edited *Media Violence and Children: A Complete Guide for Parents and Professionals* (2003, Praeger Press) and co-authored *Violent Video Game Effects on Children and Adolescents: Theory, Research, and Public Policy* (2007, Oxford University Press).

We hear conflicting reports about how video games affect our brains. One study will suggest that video games help us learn; another might imply that they make young people more aggressive. Douglas A. Gentile argues that how games influence our brains is not an either-or proposition; games can have both positive and negative consequences, and which consequences researchers find depends on what they are testing. Gentile proposes that researchers focus their investigations on five attributes of video game design to tease out these disparate effects.

VIDEO GAMERS, PARENTS, POLITICIANS and the press often lionize or attack video games, which opens the door to spin that obfuscates our understanding of how these games affect people. For example, the European Parliament has been debating whether to limit children's access to video games. In a press statement about the report that resulted from its deliberations, the parliament concluded that games could have "harmful effects on the minds of children." Reporting on this statement, however, the headline in the *Guardian* read, "Video games are good for children."

Psychologists and neuroscientists conducting well-designed studies are beginning to shed light on the actual effects of video games. These studies show a clear trend: Games have many consequences in the brain, and most are not obvious—they happen at a level that overt behaviors do not immediately reflect. Because the effects are subtle, many people think video games are simply benign entertainment.

Research projects of variable strength have substantiated claims of both beneficial and harmful effects. Too often the discussion ends there in a "good" versus "evil" battle, reminiscent of the plots of the violent video games themselves.

Games May Teach Skills—or Desensitize Us to Violence

Well-designed video games are natural teachers.[1] They provide immediate feedback on the player's success by distributing reinforcements and punishments, assist in learning at different rates and offer opportunities to practice to the point of mastery and then to automaticity. Video games also can adapt themselves to individual learners and train players in a way that helps them transfer knowledge or skills to the real world. Gamers repeat actions as they play, and repetition is one precondition for long-term potentiation—the strengthening of brain-cell connections (synapses) through repeated use that is thought to underlie memory storage and learning. To cite a mnemonic based on the theories of Canadian psychologist Donald Hebb, "Neurons that fire together wire together."

Several lines of research suggest that playing video games can lead to different types of benefits. For example, a 2002 U.S. Department of Education report presented evidence on the effectiveness of educational games.[2] One neuroscience study, published in *Nature*, showed that playing action video games can improve visual attention to the periphery of a computer screen.[3] Another study, which appeared in *Nature Neuroscience*, demonstrated that action games can improve adults' abilities to make fine discriminations among different shades of gray (called contrast sensitivity), which is important for activities such as driving at night.[4] Other research suggests that games requiring teamwork help people develop collaboration skills.[5]

Several types of studies provide evidence that video games that include "pro-social" content—situations in which characters help each other in nonviolent ways—increase such conduct outside of game play, too. In one study, 161 college students were randomly assigned to play one of several violent games, neutral games or pro-social games (in which helpful behavior was required). After playing, the students completed a task in which they could either help or hurt another student. Those who had played the violent games were more hurtful to other students, whereas those who had played the pro-social games were more helpful.[6]

Games may be beneficial for doctors, too. A study involving 33 laparoscopic surgeons—doctors who conduct minimally invasive surgery by using a video camera to project the surgical target area onto a screen as they work—linked video game play to improved surgical skill, as measured in a standardized advanced-skill training program. In fact, the surgeons' amount of game time was a better predictor of advanced surgical skill in the training drills than their number of years in practice or number of real-life surgeries performed.[7]

While some reports have linked video games to negative consequences such as obesity, attention problems, poor school performance and video game "addiction," most research has focused on the effects of violent games. Dozens of psychological studies indicate that playing violent games increases aggressive thoughts, feelings and behaviors, in both the short term[8] and the long term.[9] This makes sense from psychological and cognitive neuroscience perspectives: Humans learn what they practice. But what *really* happens in our brains when we play violent video games?

A decade ago, in an imaging study using positron emission tomography (PET), eight men undertook a goal-directed motor task for a monetary reward: They played a video game in which they moved a tank through a battlefield to destroy enemy tanks. Researchers found that a neurotransmitter called dopamine, which is involved in learning and feelings of reward, was released in the brain's striatum as the men played.[10] This and other studies suggest that the release of dopamine and stress hormones may be related not only to ideas of violence and harm, but also to motivation and winning.

Other studies have focused on how specific brain regions of players of violent games respond under varying circumstances. For instance, René Weber and his colleagues asked 13 experienced gamers to play a violent game while undergoing functional magnetic resonance imaging (fMRI) brain scans.[11] The violence in the game was not continuous, so researchers coded the game play frame by frame. At various points the player's character was fighting and killing, in imminent danger but not firing a weapon, safe with no threats, or dead.

By imaging players' brain activity before, during and after each violent encounter, the investigators found that immediately before firing a weapon, players displayed greater activity in the dorsal anterior cingulate cortex. This area involves cognitive control and planning, among other functions. While firing a weapon and shortly afterward, players showed less activity in the rostral anterior cingulate cortex (rACC) and the amygdala. Because interaction between these brain areas is associated with resolving emotional conflict, their decreased functioning could indicate a suppression of the emotional response to witnessing the results of taking violent action.

Does this evidence prove that repeated play of violent games desensitizes players to aggression and violence? In a study we are still conducting, 13 late-adolescent male gamers played a game while undergoing fMRI scans. The game (*Unreal Tournament*) can be set either to include or not to include violent actions. The most interesting preliminary findings appear in the contrast between gamers who habitually play violent, first-person shooter games and those who play less violent games. The latter show increases in rACC activity (suggesting emotional responses) during violent episodes, as expected. We interpret this to mean that while people

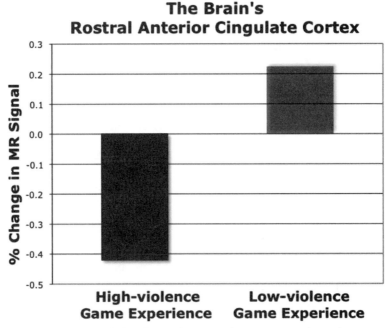

Figure 1 When playing a high-violence video game, players accustomed to such games showed lower activity (measured via signals from magnetic resonance imaging) in the rostral anterior cingulate cortex (rACC), whereas players used to low-violence games displayed higher activity. This difference suggests that gamers who often play violent games may be desensitized to aggression and violence. (Courtesy of K. Thomas and D. A. Gentile)

who are not used to seeing violent images show a strong emotional reaction when confronted with them, those who regularly play violent games do not simply lack an emotional reaction—they actively *suppress* it, as reflected in their rACC activity (see Figure 1). These players may have become desensitized to violence in video games. Another study has found a correlation between repeated exposure to violent games and desensitization, as well as increased aggressive behavior in the real world.[12]

Researchers are continuing to investigate whether repeated exposure to violent games over time truly does desensitize players and increase their aggressive feelings and thoughts. Beyond that, we must find out whether such responses and behaviors become automatic.

Aspects of Video Games Illuminate Their Effects

If video games can be both beneficial and harmful to players, how can we predict their effects on individuals and populations? And how can scientists stay out of the polarizing debate while still reporting the results of their studies? The answer to the second question is that scientists must be precise. I see five aspects of video games that can affect players: amount, content, structure, mechanics and context. Together, these aspects can explain different research results.

Amount

We would expect that as people spend more time playing video games, their risk of performing poorly in school, becoming overweight or obese, and developing specific negative physical health outcomes (such as carpal tunnel syndrome and other repetitive stress injuries) would increase. We may also correlate more time spent playing with a higher number of video-induced seizures in people with epilepsy or photosensitivity disorder.

These correlations might begin with gamers' existing characteristics. For example, low-performing students are more likely to spend more time playing video games, which may give them a sense of mastery that eludes them at school. Nonetheless, every hour that a child spends on video games is not spent doing homework, reading, creating or participating in other activities that might have more educational benefit. Longitudinal studies support the idea that children's school performance worsens as their gaming time increases.

Furthermore, excessive video game play often reduces time for physical activity, which could account for the link between large amounts of gaming and obesity. Movement games (such as *Dance Dance Revolution* and some Nintendo Wii games) may have the opposite effect, however. Finally, repetition of a game's features may magnify the consequences of the four other aspects I cite.

Content

What a video game is about—its content—may determine what players take with them from the game to real life. Studies indicate not only that games that include educational content can improve related education skills,[2] but also that games designed to help children manage chronic health problems (such as asthma or diabetes) are more effective than doctors' pamphlets in training children to recognize symptoms and to take their medications.[13] Similarly, studies reporting that games with violent content increase aggressive thoughts, feelings and behaviors suggest that these violent tendencies can extend into real-life situations.[6]

Learning that results from video games may last only for the duration of the game, for a few minutes after play ceases or for the long term. Many content-focused studies, such as those in which children learn information about their health, also show that in-game learning can transfer to the real world in the long term.

Studies of games with violent content also tend to demonstrate transfer of learning to real-world situations. Studies in several countries, including one consisting of 1,595 children in Japan and the United States, suggest that children who play violent games become more aggressive in their daily lives (as reported by their peers or teachers, for example).[9]

The research question has shifted from *whether* game content transfers to nongame situations to *how* it does so. New studies are focusing on the cognitive and behavioral processes by which learning and transfer occur. Learning may transfer to other tasks as a result of fluid intelligence. This term refers to the ability to reason and to solve new problems independently of previously acquired knowledge. In one study, researchers had participants complete a gamelike computer training task. They found that training designed to improve working memory (the temporary storage and use of information) leads to a transfer to fluid intelligence. Moreover, the extent of the gain was related to the amount of training.[14]

Structure

A video game's on-screen structure contributes to its effects. For example, some games require a player to scan the screen constantly for small changes, such as signals announcing the sudden appearance of an "enemy," and to respond quickly to these changes. Effective scanning allows the player to shift attention swiftly and automatically from the center of the screen to the periphery.[3] Such visual attention is analogous to the type of skill that an air-traffic controller needs: the ability to scan all screen areas, to detect minute changes and to respond quickly.

Some games require players to navigate three-dimensional virtual worlds on a two-dimensional screen. To navigate successfully, players must use multiple depth cues (such as interposition, in which closer objects obscure more distant ones, and motion parallax, in which objects move across the visual field faster if they are closer to the viewer). These games should improve a gamer's ability to get 3-D information from 2-D depth cues and use it in other contexts. Because navigating a virtual world requires players to maintain awareness of orientation and landmarks, these games also could improve way-finding skills and mental rotation skills. Such transfer could explain the findings of the laparoscopic surgeon study, because the surgeons need to gather detailed 3-D information from a 2-D screen while maintaining awareness of both the screen's periphery and objects that are *not* on the screen. The biological basis of how a game's structure affects players requires further research.

Mechanics

We should anticipate that the mechanics of game play require gamers to hone particular motor skills, which may also transfer to related real-life situations. A game's mechanics relate to its structure. Movements of the controller change what a player sees on the screen, which in turn affects how the player uses the controller. This feedback loop is consistent with hand-eye coordination.

We would expect to see improvements when gamers practice isolated movements or coordinate them between dominant and nondominant hands, in terms of both fine motor skills (such as making tiny adjustments with the thumb on a game pad) and gross motor abilities (such as practicing a baseball swing on the Nintendo Wii). This expectation also applies to the use of the devices required to play a game. For example, playing driving games with a wheel and pedals is likely to transfer to real-world driving more effectively than playing the same game with a keyboard and mouse.

Context

Finally, we should suppose that the social context of a game influences its effects on the brain and learning. Some games require cooperation and teamwork for success. For example, in some quests of the multiplayer online game *World of Warcraft*, players with different skills must work together to solve puzzles and to overcome barriers. Other games, such as the battle simulator *Call of Duty*, require real-time coordinated action. Games that involve teamwork may improve players' skills in cooperation and coordination, but scientists have conducted almost no research in this area.

A game's social context may change other outcomes as well. For example, playing violent games with a group of friends who provide social support for aggressive actions might yield greater increases in aggressive behaviors in other contexts than playing the same games by oneself. Conversely, providing players with pro-social motives to help their friends might mitigate increases in aggressive behaviors. Because social context is more difficult to study than the consequences of game content or amount of play, researchers will need to design new experimental methods.

Examining these five aspects of video games has several benefits. It allows us to get beyond the dichotomous thinking of games as simply good or bad. It helps us understand why different types of studies have different outcomes. Finally, it tells us why these findings do not actually contradict each other but simply represent different levels of analysis.

With the exception of educational games, most video games' effects on brain and behavior are unintentional on the part of both the designers and the players. Nonetheless, research suggests that the effects are real. Video games are neither good nor bad. Rather, they are a powerful form of entertainment that does what good entertainment is supposed to do—it influences us.

7

Updating the *Diagnostic and Statistical Manual of Mental Disorders*

By David J. Kupfer, M.D., Emily A. Kuhl, Ph.D.,
William E. Narrow, M.D., M.P.H.,
and Darrel A. Regier, M.D., M.P.H.;
by Paul R. McHugh, M.D.

David J. Kupfer, M.D., is Thomas Detre Professor and chair of the department of psychiatry at the University of Pittsburgh and the Western Psychiatric Institute & Clinic. He is also chair of the DSM-V Task Force.

Emily A. Kuhl, Ph.D., is the science writer of the American Psychiatric Institute for Research and Education at the American Psychiatric Association.

William E. Narrow, M.D., M.P.H., is associate director of the division of research at the American Psychiatric Association. He is also research director of the DSM-V Task Force.

Darrel A. Regier, M.D., M.P.H., is executive director of the American Psychiatric Institute for Research and Education and director of the division of research at the American Psychiatric Association. He is also vice chair of the DSM-V Task Force.

Paul R. McHugh, M.D., is University Distinguished Service Professor of Psychiatry at the Johns Hopkins University School of Medicine. From 1975 to 2001, he was the director of the department of psychiatry and behavioral science at the Johns Hopkins School of Medicine and psychiatrist-in-chief at the Johns Hopkins Hospital. He is the author or co-author of five books and has published more than 200 articles in scientific journals.

The Diagnostic and Statistical Manual of Mental Disorders *(DSM), which psychiatrists and other practitioners use as a guide to diagnose psychiatric disorders, is in the early stage of revision, to be released in 2013. The manual has long been controversial; starting with the third edition, published in 1980, it has focused on using symptoms to diagnose disorders and has purposely avoided speculating on their causes. In the first of these complementary articles, scientists leading the DSM revision process explain how they are attempting to bring both more certainty and more flexibility to psychiatric diagnosis. In the second article, Johns Hopkins psychiatry professor Paul R. McHugh strongly urges DSM's editors to focus on disorders' causes and disease processes and to improve upon what the two most recent editions of DSM have produced: "a psychiatry that's boring."*

On the Road to DSM-V

By David J. Kupfer, M.D.; Emily A. Kuhl, Ph.D.; William E. Narrow, M.D., M.P.H.; and Darrel A. Regier, M.D., M.P.H.

The *Diagnostic and Statistical Manual of Mental Disorders* (DSM), first published in 1952 and considered the foremost text for mental health specialists to recognize and diagnose mental illnesses, is undergoing revision. Almost 15 years' worth of research and scientific advances since the fourth edition of the manual (DSM-IV) was released has made it evident that the system used by tens of thousands of psychiatrists, psychologists, social workers, primary care physicians and psychiatric researchers does not adequately mirror what they see in the clinic or laboratory.

The manual provides an index of psychiatric disorders (such as anxiety, mood and eating disorders) categorized by their core symptoms. Disorders are paired with numerical codes, based on the International Classification of Diseases, that are entered into the patient's medical record and used for medical record keeping, reimbursement from insurance companies and to help researchers compile statistical data. Each listing includes a description of the illness and its features, followed by a listing of symptom criteria required for diagnosis. Clinicians use the DSM's system of five diagnostic "axes" to evaluate the condition of interest as well as the presence of personality disorders and mental retardation, psychosocial and environmental problems (such as marriage difficulties or the presence of physical abuse) and general medical conditions. The fifth axis allows for a rating of how a patient is functioning overall (such as at work or at school).

Although DSM is the pre-eminent resource for psychiatric diagnosis in this country, it does have flaws. For example, DSM-IV categorizes disorders by shared features or symptoms; it classifies major depressive disorder and bipolar disorder as mood disorders because they have several overlapping diagnostic criteria. But people may show signs of multiple syndromes, their symptoms may range from mild to debilitating, and some people have unique presentations that do not fit the diagnostic mold. The manual's system accounts for none of these situations.

DSM-IV, like its predecessors, fails to capture the diversity and complexity within disorders that make them both fascinating and perplexing to understand and to treat. Advances in neuroscience, epidemiological data and evidence from clinical care all underscore the fact that psychiatric diagnoses are not neatly compartmentalized entities with clearly defined boundaries. Although other areas of general medicine have their own ambiguities and difficulties, they convey a greater sense of certainty through diagnostic tests, laboratory values (such as levels of blood glucose or cholesterol), radiology scans and other quantitative measurements. Psychiatric states are not as reliably measured.

While such ambiguous boundaries make for enriching and challenging detective work for psychiatric clinicians and researchers, they do not lend themselves to an efficient method of organizing and conceptualizing diagnoses. Even the most seasoned psychiatrists apply DSM-IV guidelines with some level of uncertainty, and non-psychiatric professionals—who frequently have little training in psychiatric disorders and even less time to conduct a thorough clinical evaluation—are at an even greater disadvantage. Neurobiology, imaging and genetics are providing a more tangible, quantifiable understanding of psychiatric disease, but the state of the science in the early 1990s, when DSM–IV was released, did not allow its editors to incorporate these advances. Because DSM remains our primary source for diagnosis, the new edition must be able to reference the rapidly emerging scientific research and incorporate such findings when the empirical foundation supports them.

The Development of DSM

To understand why DSM-IV fails to represent the complex nature of psychiatric disorders, such as the commonly seen mixture of anxiety and depression, we first need to understand the evolution of our psychiatric diagnostic system.

In the late 1960s (the era of DSM-II), disagreement between clinicians and researchers about how to diagnose and treat psychiatric disorders correctly was growing. In response, a group of psychiatrists at Washington University in St. Louis set out to develop a revised set of diagnostic criteria based on a review of nearly 1,000 published articles and existing data. Project leaders Eli Robins, M.D., and Samuel Guze, M.D., allowed one of their young residents, John Feighner, M.D., to be the first author on a paper published in 1972.[1] The paper discussed 15 disorders whose descriptions were based on criteria that the authors believed could be corroborated by future research, thereby enhancing the validity or legitimacy of those criteria. Two years earlier, Robins and Guze had published a set of validity criteria that practitioners could use to test diagnoses.[2] Under these criteria, disorders are considered valid if they separate clearly from others, follow a predictable clinical course, aggregate in families and eventually have distinct laboratory tests. The "Feighner criteria" introduced in the 1972 paper explicitly identified disease symptoms and durations, a stark contrast to the vague descriptors of DSM-II. The detailed and explicit nature of the criteria made it possible for clinicians to identify similar symptom patterns in patients in different settings, thereby increasing the consistency and reliability of psychiatric diagnosis.

Shortly thereafter, the National Institute of Mental Health (NIMH) initiated the Collaborative Psychobiology of Depression Study. The participating investigators wanted to ensure that all patients who were to be enrolled in the study had the same essential syndromes. Robins, the Washington University psychiatry department chair and one of the study's lead investigators, expanded the Feighner criteria to create the Research Diagnostic Criteria (RDC).[3] The investigators developed a standardized patient interview tool called the Schedule for Affective Disorders and Schizophrenia (SADS)[4] to be sure that patients who entered the NIMH study truly met Robins' criteria.

The RDC became the basis for DSM-III, radically shifting the method of diagnosis from a system that used Freudian theories of causation to one based on organizing diseases according to similarities in symptoms and their duration. (DSM-IV continued this system.) For instance, the classification of anxiety in DSM-II is subdivided into forms of neuroses, including anxious, hysterical and hypochondrical, each accompanied by a text description that describes symptoms but without identifying explicit criteria (hysterical neurosis, for example, is "characterized by an involuntary psychogenic loss or disorder of function"). Using the RDC, the new approach in DSM-III provided explicit symptom and duration criteria for each disorder without implying either Freudian or biologically based theories of cause.

DSM-III showed that greater reliance on explicit criteria drastically improved diagnoses' dependability and consistency. However, it introduced a system in which a "higher-order" disorder subsumed all "lower-order" disorders in the following hierarchy: traumatic/infections brain diseases, schizophrenia, manic-depressive disorder, major depression, anxiety disorders, somatization disorder (multiple unexplained physical symptoms), substance use and personality disorders. Under this system, patients could not be simultaneously diagnosed with both a higher-order and lower-order disorder, as the hierarchy implied that symptoms of the lower-order disorder would also be found in the dominant disorder, making a dual diagnosis unnecessary and redundant. Immediately after the release of DSM-III, a large, NIMH-supported epidemiological study used DSM-III diagnostic criteria to identify prevalence rates of mental disorders in community, hospital and institutionalized populations.[5] This study demonstrated that a strict implementation of the proposed hierarchical restrictions would suppress a great deal of descriptive clinical information because most individuals who met criteria for one disorder also did so for a second or third—but only one could be diagnosed.[6] This finding suggested that the hierarchical approach obscures the true complexity of some psychiatric disorders and, by obfuscating important targets for clinical research, could be hindering the development of appropriate treatments. The revised DSM-III (DSM-III-R) partially abandoned this hierarchy but resulted in a large number of patients diagnosed with multiple disorders—a problem that persists in DSM-IV.

Treating patients who have multiple psychiatric diagnoses poses a significant challenge. For example, recent analyses[7] have shown that people with non-psychotic major depression occurring alongside anxiety or substance use did not respond as well to treatment as those with depression alone. A majority of the major-depression patients in these analyses exhibited such co-occurrences, including about 10 percent who had both anxiety and substance use disorders. In addition, a recent study of primary care patients who had reported severe depression, anxiety or somatization disorder revealed that more than half had been diagnosed with at least two of these.[8] Patients with all three disorders had significantly more difficulty maintaining physical health and social relationships, adding to the complexity of finding the most appropriate treatment for them.

Despite these and other findings suggesting that "pure" disorders are rare, the DSM-III classifications describe such disorders, easily distinguishable from one another and from healthy behaviors.[9] The structure of DSM-IV perpetuates this misperception. Further, DSM-IV has not entirely abandoned the DSM-III hierarchy; for example, a psychiatrist following the criteria today can diagnose neither attention-deficit/hyperactivity disorder in the presence of autism nor generalized anxiety disorder if it occurs exclusively in the presence of a mood disorder.

DSM-IV compartmentalizes diagnoses into strict categories that do not reflect the most common symptom patterns that actually appear in patients. For example, the criteria for Major Depression fail to reflect the potential co-occurrence of anxiety symptoms, which appear in more than 50 percent of patients with depression. If a patient fails to meet full criteria for Major Depression and has significant anxiety symptoms, which together cause significant distress and impairment, the diagnosis would fall under the "not otherwise specified" (NOS) category. People who receive such diagnoses do not officially meet criteria for any specific DSM disorder, yet their symptoms may be severe and they may have so much difficulty with ordinary relationships and daily activities that they warrant attention and possibly treatment. Conditions that fall just short of diagnostic requirements, mixed disorders, and those with uncommon or unusual symptoms all may land in the NOS category. Because clinical trials of psychotherapy or medication are conducted only for DSM-defined disorders, an NOS diagnosis makes it difficult for a physician to choose an appropriate evidence-based treatment.

The Utility of Diagnostic Dimensions

One of the more promising pathways out of the categorical conundrum that the DSM revision task force is addressing is a "dimensional" approach—one that enables clinicians to consider distinctive aspects that differ significantly within a disorder (such as symptom severity), as well as the presence of symptoms that are outside the "pure" disorder definitions (such as anxiety and somatic symptom levels for patients with depression). This method incorporates variations of features within a disorder (its severity and whether it is acute or chronic, for example) rather than relying on the answers to simple yes or no questions to arrive at a diagnosis. Dimensions also can be used to examine features of other diagnoses. For example, if DSM-V provided for clinicians' documentation of certain symptom dimensions in all patients—such as sleep/wake functioning, cognition, mood and anxiety symptoms, substance use and psychosis—the result would be a more useful and realistic representation of the patients' clinical status than that of the current method. The dimensional approach also helps reduce the need for multiple diagnoses, provides background explanation for an NOS diagnosis, clarifies the presence and severity of individual symptoms and informs treatment planning.

Capturing variations would increase DSM-V's clinical utility and support a systematic, measurement-based approach to monitoring symptoms and their severity when making decisions about treatment.[10]

Researchers already have shown such measurement-based care approaches to be both feasible and useful in primary care and mental health settings.[11] For instance, investigators found that psychiatrists readily accepted and used one such dimensional measure, the nine-item depression scale of the Patient Health Questionnaire (PHQ-9). Further, psychiatrists reported that they used results of the questionnaire when making treatment decisions for patients with major depression during approximately 40 percent of patient visits. These decisions included changing the dosage of an antidepressant; adding, introducing or switching antidepressants; initiating or increasing psychotherapy; and engaging in additional suicide risk assessments.[11] Although practitioners can certainly make such treatment decisions without guidance from measurement-based care methods, these dimensional assessments provide objective data that guide clinical decisions and that might otherwise be overlooked in busy, real-life clinical settings. Notably, 42 percent of the patients in the sample also had an anxiety, substance use or other psychiatric disorder. This indicates that even in real-world settings, where patients' symptoms vary and time is at a premium, dimensional assessments can assist in diagnosing disorders as well as measuring severity, predicting outcomes and planning treatment.

The expectation that the Robins and Guze validity criteria would lead to the validation of disorders as distinct entities has gone largely unfulfilled. For example, recent advances in neuroscience demonstrate that the classification structure of DSM-IV that strictly separates schizophrenia and bipolar disorder may not be scientifically valid. Family, twin and adoption studies have repeatedly demonstrated an overlap in many genetic markers associated with a susceptibility to these two disorders.[12] This parallel suggests a possible shared genetic vulnerability to some symptom patterns that belies these disorders' current representation as diagnostically distinct entities. Although they have some notable clinical differences, their strict separation in DSM-IV only encourages clinicians and researchers to persist in conceptualizing them as fundamentally discrete. The use of dimensions also likely will shed light on how other now-separate disorders with shared multiple-gene vulnerabilities and shared symptoms have a tendency to cluster, thus raising important questions about what the overarching structure of DSM-V should be.[13,14]

These multiple-gene susceptibility findings lend further support to a reorganization of DSM that moves away from a strict, categorical, "yes/no" approach that was more consistent with the previously prevailing but now obsolete idea that most mental disorders could be linked to a single gene. Basing DSM in part on findings from neurobiological studies is one proposal. For example, studies of genetic variations within groups of people with schizophrenia and of twins where only one sibling has schizophrenia indicate that this disorder arises from a complex combination of genetic and environmental factors.[15] People with schizophrenia may have relatives who do not meet DSM-IV diagnostic criteria for the disease but nonetheless exhibit abnormalities consistent with it—both neurobiological (such as thinning of the brain's cortex) and neuropsychiatric (such as impaired cognitive abilities).[12] The presence of overlapping conditions, such as bipolar disorder and schizophrenia—or depression and anxiety—has led many experts to suggest that concurrent symptom patterns that cross existing diagnostic boundaries may constitute more aptly named syndromes, such as "affective-laden schizophrenia" or "anxious depression."

To reorganize categories in DSM-V to better reflect the state of the science, we will have to do more than simply revise individual diagnostic criteria in DSM-IV. Realistically, given the state of the science, psychiatry as a field and of DSM users in general are not yet ready for a drastic overhaul of DSM's organization. As such, DSM-V revision experts are examining whether specific indicators can inform and validate the grouping of disorders while maintaining much of the existing categorical framework.

Eleven such indicators are currently under examination. Indicators[16] shared by multiple disorders include:

- neural substrates (brain structures that underlie a behavior or psychiatric state),

- genetic risk factors,
- specific environmental risk factors,
- biomarkers (such as a feature on a brain scan that denotes disease presence or progression),
- temperamental antecedents (characteristics of temperament that increase risk for a particular disorder),
- abnormalities of emotional or cognitive processing,
- treatment response,
- familiality (common occurrence within a family),
- symptom similarity,
- course of illness
- and high comorbidity (the co-occurrence of two or more conditions).

These indicators can serve as evidence-based guidelines to inform decision-making about how to cluster disorders to maximize their validity and therefore their utility for clinicians.

Regrouping psychiatric disorders will enable future researchers to enhance our understanding of the origins and common disease processes (pathologies) among disorders. It will also provide a base for future changes that reflect advances in the underlying science: Data can be re-analyzed over time to continually assess the groupings' validity. Thus, after DSM-V is published, changes to the volume will occur only to the extent that future discoveries in neurobiology, genetics, epidemiology and clinical research support them.

One Vision: A Psychiatric "Review of Systems"

To advance clinical practice and to provide a framework for future testing of the standards for diagnosing mental disorders, the forthcoming DSM-V criteria need to better reflect the true nature and scientific underpinnings of psychiatric disorders while preserving their link to previous diagnostic conventions.[10] An important strategy for achieving these objectives involves the integration of previously described dimensional measures with the current criteria that define mental disorders. By recommending patient self-report screening methods that cut across multiple diagnostic areas, the DSM-V will facilitate a more systematic review of multiple symptom domains (brain or mental systems). This approach is comparable to general medicine's review of systems, which resembles casting a fishing net that simultaneously captures everything at once and nothing in particular. In general medicine, this broad review process is crucial for detecting pathological changes in different organ systems when creating a comprehensive diagnosis and treatment plan.

Similarly, DSM-V will provide recommendations for guiding practitioners through a systematic review of multiple brain or mental systems to prevent a premature focus on the most obvious symptoms—thereby helping mental health practitioners conduct a more comprehensive mental status assessment. When an initial screening reveals one or more symptoms in different domains, patients and clinicians will be directed to follow up with a more intensive dimensional evaluation of symptom severity, followed by a clinical judgment about whether the symptoms are sufficient for a mental disorder diagnosis. This combined dimensional and categorical diagnostic approach will also help clinicians and research investigators establish a database of better-classified syndromes for future clinical, epidemiological and neurobiological study.

Even though the DSM comprises diagnoses, we must remember that its construction is not so much about pathology as it is about people. Our aim with DSM-V, first and foremost, is to improve patient care. The unique features that a patient brings to an assessment—family background, life experiences, social functioning and relationship history—are as important as the symptoms themselves; without the relevant personal information, a physician observing symptoms alone may not make a correct diagnosis. The science behind DSM-V should therefore serve to strengthen, not to overshadow, clinical care by connecting the most recent scientific findings to the objective information each clinician and patient brings to diagnosis and treatment. If, for example, Tourette's syndrome shares observable symptoms and underlying biomarkers with obsessive-compulsive disorder, clustering the two in DSM would encourage clinicians to look for tics, a common symptom in Tourette's but not in obsessive-compulsive disorder.

Research gains in recent years will advance the scientific validity and clinical utility of DSM-V, scheduled for publication in May 2013. As new findings from neuroscience, imaging, genetics and studies of clinical course and treatment response emerge, the definitions and boundaries of disorders will change. Perhaps the most important characteristic of DSM-V is that it will be a living document with a support system for a continuous review and revision process.

Psychiatry at Stalemate

By Paul R. McHugh, M.D.

The Diagnostic and Statistical Manual of Mental Disorders (DSM) of the American Psychiatric Association (APA) has, from its first edition (DSM-I) of 1952, represented the official taxonomic enterprise of American psychiatry: identifying, delimiting and classifying the diverse mental and behavioral disorders that psychiatrists aim to treat or prevent. Although the series was launched primarily to collect statistical information on mental disorders, with its third edition (DSM-III), published in 1980, this previously descriptive enterprise took a new and prescriptive turn and began directing psychiatric diagnostic practice. Serious challenges to the legitimacy of the discipline—many from within the profession itself— provoked this new effort. They ranged from claims that disorders were not properly differentiated or aptly treated to assertions that mental illnesses were social fabrications of psychiatrists—"myths," in fact. None of these accusations could be brushed aside given that, at the time, diag- nostic agreement between two psychiatrists about the same patient was hardly better than chance.[1]

In the face of such foundational disarray, Robert Spitzer, the editor of DSM-III, proposed to make psychiatrists more "scientific" (and thus more medically legitimate) by prescribing for them, through the new edition, a better way of identifying and distinguishing mental disor- ders. Psychiatrists, he said, should use what he dubbed an "operational," "a-theoretical" method that could, by identifying each mental disorder from its symptomatic presentation, steer clear of the conflicts over those explanatory theories of mind dividing psychodynamic and neurobiologically oriented psychiatrists on which much of the disarray rested. With awesome diplomatic and political skill Spitzer persuaded the profession to accept this new approach, and now for a generation American psychiatrists have employed DSM's symptom-based method in practice, teaching and research. DSM-IV, published in 1994 with minor changes, followed the same path, so that hereafter I shall refer to the pair as DSM-III/IV.

DSM-III/IV helped resolve the turmoil that fractured psychiatric discourse in the 1970s by getting all, psychoanalysts as well as neurobiologists, to concentrate on the few things they could agree about: the symptoms patients displayed and how those symptoms naturally emerged over time. This approach succeeded so well that arguing today about "dynamic" and "biological" psychiatry is an anachronism. And no one seriously suggests that mental disorders are "myths," especially given that the diagnostic consistency of DSM-III/IV improved the efficacy of both psychological and pharmacological treatments.

And yet all is not well with psychiatry under this new dispensation. Many difficulties have emerged directly from it; they are serious enough to challenge the usefulness of any revision of DSM that does not make a significant move to resolve them.

Most of these problems derive from the ad hoc nature of DSM-III/IV and its glorification of process over substance. It aims only to enhance diagnostic consistency. It does not speak to the nature of mental disorders or distinguish them by anything more essential than their clinical appearance. Not a gesture does it make toward the etiopathic principles of cause and mechanism that organize medical classifications and carry, for physicians and patients alike, the promise that rational, effective, cause-attacking therapies will eventually replace symptom-focused, palliating ones.

I and others contend that this symptom-based approach, having accomplished its original purpose of settling the discord within psychiatry, should now gradually but resolutely be supplanted. What we needed in 1980 is not what we need now, a generation later. In fact, today's pressing issues are those produced by DSM-III/IV.

The most fundamental of these problems is that DSM regularly fails to distinguish between conditions with similar symptomatic appearances such as between ordinary sadness and clinical depression, as Allan Horowitz and Jerome Wakefield have recently and thoroughly documented.[2] This failure derives directly from the inattention of DSM-III/IV to distinguishing the generative causes of either normal or abnormal mental states.

But the method and purposes of DSM are so aligned (and the profession so accustomed to them) that practitioners and editors alike resist suggestions for revising a new edition in ways more substantial than tinkering with the criteria and expanding the collection of certified conditions. Psychiatry has become a field bridled by its own method and needs to fight its way free.

A Classificatory Dead End

The symptom-based diagnostic method of DSM differs fundamentally from the way doctors classify and distinguish medical disorders. Internists differentiate disorders according to how they display—through their symptoms, signs and laboratory data—some functional disruption of bodily design, such as the pump function of the heart in congestive heart failure, the filtering function of the kidney in renal diseases or the gas-exchange capacities of the lungs in pulmonary disease.

Medical classifications of this sort—properly identified as generative in that they build upon concepts of cause or mechanism generating the conditions—evolve and improve over time, not simply because they follow progress in the natural sciences (physiology in particular), but crucially because they interact with and enrich such sciences with information from physicians who recognize diseases as "experiments of nature" revealing of mechanisms behind the symptoms and their course. For example, William Harvey not only used the experimental method with animals to demonstrate the circulation of the blood but also pointed to the features of human congestive heart failure to demonstrate just what symptoms and signs appear when that circulation begins to falter. Because this scientific partnership between medicine and biology became so successful, time-honored but historically separate modes of thought—the "healing" tradition of medicine and the "natural history" tradition of biology—today merge as aspects of "life science," to our benefit.

No symptom-based, appearance-driven diagnostic system such as DSM can do the same for psychiatry and neuroscience. It is too passive in relation to knowledge; it awaits discovery rather than suggesting approaches to it; it remains satisfied with consistent analytic definitions of mental disorders at the expense of any synthetic grasp of them from their origins and generation. In all these ways DSM-III/IV has produced what no one ever thought possible—a psychiatry that's boring.

A close look at its actual classificatory operations reveals why, resistant to change, DSM has reached a dead end.

DSM-III/IV, like any way of classifying things, must deal with two issues in confronting their diversity—here, of the mental disorders. The first is the demarcation, definitional issue whereby individual disorders are identified. The second is the separation, differentiation issue whereby the defined disorders are separated from one another. DSM-III/IV insists on addressing both of these issues "a-theoretically"—that is by an approach that attempts to be "neutral with respect to theories of etiology" (DSM-IV p. xviii).

DSM-III's a-theoretical approach to the demarcation/definitional issue was simple; it "asked the experts." Spitzer selected psychiatrists he believed to be expert students of each disorder and solicited definitions and diagnostic rules from them. He, with his fellow editors, refined these suggestions and chose for the official diagnostic algorithms the symptoms easiest to elicit from patients in an interview and to use with consistency in practice.

This definitional exercise now tends to go the other way. "Experts"—many unfortunately with a vested interest (financial, political, legal, ideological) in gaining an official "stamp" certifying the existence of a particular mental condition—now beat on DSM's editorial door for the inclusion of their favorite malady in the manual. Because no more objective criterion than clinical testimony can be employed to challenge an admission to the DSM catalog, these "experts" cannot be denied if they are a sizeable "lobby" and bring with them a set of user-friendly diagnostic symptoms for the condition they want listed.

This a-theoretical approach to the demarcation/definition issue has produced several problems. First, some DSM conditions are defined with such inclusive symptomatic criteria that they gather in too many patients and so are diagnostically overused to the detriment of refined assessment and treatment. Such DSM-III/IV diagnoses as Bipolar Disorder and Attention Deficit Disorder, for instance, are now so frequently applied, and so many people are inappropriately receiving Prozac or Ritalin, that these diagnoses and medications have become household words to the American public and material for late-night television comedians.

Second, many varieties of the same disorder are separated in the DSM because it emphasizes trivial distinctions in symptom expression. For example, patients given diagnoses of narcissistic personality disorder, histrionic personality disorder, or borderline personality disorder are all unstable ('high neuroticism") extraverts who tend to be disagreeable. The specific diagnostic label they receive depends more on what feature a doctor chooses to emphasize than upon anything psychologically distinct or critical to their treatment. A patient who seeks a second opinion across town may well receive one of the other labels—and it would be just as correct, according to the DSM.

Finally, because DSM lacks any other way of judging what fits as a legitimate psychiatric condition but must accept what "experts" champion, it grows in size with each edition, becoming ever more impressive in its list of diagnoses even as it remains ever so humble in its explanations of them.

DSM-III/IV also manages the separation/differentiation issue of classification a-theoretically. The editors employ an old and familiar method of drawing distinctions as demonstrated by the decision trees in Appendix A of both DSM-III and IV. These depict the traditional "downward" method for dividing large classes into progressively smaller sets by using a sequence of dichotomizing (yes/no) questions. For example, the decision tree recommended for the specific diagnosis of an anxious patient uses a sequence of such dichotomizing questions as "due to medical condition," "cued by situation" or "response to traumatic event" until a yes picks out, say, generalized anxiety disorder or acute stress disorder as fitting the patient of interest.

This method is traditional in being first formally described by Aristotle. But it resembles nothing so much as the children's "Twenty Questions" game, wherein a player, by means of a sequence of yes/no questions ("animal or mineral," "warm blooded or cold blooded," "feathered or furred" and so on), ultimately identifies the object the other player has in mind. It also is the standard method of naturalists' field guides such as Roger Tory Peterson's *Field Guide to Birds of North America*. It works when the sole aim is identification and when the dichotomizing questions are easy enough to answer in the field, as Peterson's many fans will testify.

But clinicians have several problems with this method of demarcation. Because the dichotomizing questions that ultimately determine a diagnosis are to some extent arbitrary, the method is vulnerable to abuse when advocates interested in producing a given result devise a way of inserting their own distinctions in the sequence. The best example of this is the artificial distinction drawn in the DSM between "conversion disorders" (where patients display pseudo-neurological phenomena of a physical kind, such as paralysis or sensory loss) and "dissociative disorders" (where patients display pseudo-neurological phenomena of a psychological kind, such as amnesia or multiple personalities). A diagnostic distinction between these two expressions of illness-imitating, attention-seeking behavior implies that they are different in some essential way. In reality they are behaviors of the same nature in that both are provoked by suggestion, display symptoms that can attract contemporary clinical attention and services, and, not that infrequently, are exhibited by the same patient.

Another critical problem is that this downward method of differentiation in psychiatry operates with so little information—certainly none of a psychological or neuroscientific kind—that it confounds those symptomatic expressions that are primary and essential to a disorder with those that are secondary and adaptive, such as the depressive and paranoid reactions shared by many disorders. The method hides this diagnostic and therapeutic complication by emphasizing the consistency of its final decisions.

The result is the situation we have today. A process aimed at producing diagnostic consistency has not only generated several practical problems of its own but has reached a dead end where the only route of escape is the one that method categorically rejects: the re-introduction of concepts of cause and mechanism—theories, in fact—into the diagnostic reasoning of the discipline.

However, at a meeting of the Johns Hopkins Department of Psychiatry in 2008, Michael First of Columbia University, who has had senior editorial responsibility for DSM, told us that the editors all agree that despite the increase in psychiatric research that followed the publication of DSM-III in 1980, nothing has emerged in the 30 years since that permits us to diagnose *any* condition in the DSM by the medically traditional etiopathic—cause or mechanism—approach. Thirty years with a field guide and nothing on the horizon offering another way. And, yet the editors of DSM-V say it must come forth as "Son of DSM-IV."

Surely one can wonder about the wisdom of this advice. Can DSM-V offer us nothing to provide a better conceptual grasp of mental disorders or, at a minimum, suggest—in the form of reasonable hypotheses based on psychological and neuroscientific evidence—their nature, mechanism or cause?

Just consider the point of view of a patient who has received a DSM-III/IV diagnosis. What does he take away on learning from his doctors that his distressful state of mind "satisfies criteria" for Major Depression? Should he presume that he is afflicted with a disease—something he *has*—or should he think of his problem as an emotional state or reaction to something he *encountered*? Or should he strive to realize that the problem is a propensity for low spirits tied to his personality—something he *is*—or should he consider it a state of mind produced by how he is behaving—something he is *doing*? The diagnostic label he has received makes none of this clear to him.

It's all so tautologous. You're miserable, seek medical attention and are told you've got Major Depression! How much more do you know?

And what's the most rational treatment now that you have a label? From where is the help to come?

A Modest Suggestion

Official psychiatry is at stalemate. It must produce a new edition (DSM-V) soon to fit the World Health Organization's schedule for updating the International Classification of Diseases (ICD), used worldwide for diagnostic and clinical purposes, and for epidemiological studies of disease prevalence and death rates. But currently, most revision proposals either amount to little more than tinkering within the DSM symptom-based diagnostic system or are too uncertain of outcome to be encouraged. Many psychiatrists fear that any classificatory differentiation based on views about the generation of psychiatric disorders will restart the war between the dynamic and biological schools that DSM-III settled.

But surely DSM-V's editors can take some tentative steps toward classifying psychiatric conditions by what underlies them—particularly if these steps are based on modes of thought ever implicit in much of psychiatric practice and research. Simply making explicit what has been implicit would be progress.

A grouping of disorders, not by their symptomatic similarity but in families that share a causal, generative nature, would introduce into DSM the etiopathic principle fundamental to medical classifications. The DSM already employs something rather like this idea in that it separates disorders that derive from intelligence and temperament into its "Axis II" grouping and so distinguishes them from all other disorders, which it clusters on "Axis I."

The discipline would surely advance if DSM specifically separated those disorders that represent *breakdowns in* the mind's design and indicate brain disease (such as dementia, delirium and schizophrenia) from those that represent disturbed *expressions of* the mind's design in the form either of behavioral misdirections (anorexia nervosa, alcoholism and sexual paraphilias) or emotional responses to distressful life encounters (adjustment disorder, post-traumatic stress disorder and bereavement).

Such a reorganization of the catalogue would not require abandoning the familiar DSM-III/IV diagnostic algorithms. Rather, it could simply be superimposed upon them. And so, consistency of diagnosis would be retained even as the possibility of eventually resting diagnoses upon generative mechanisms would be foreshadowed.

By encouraging clinicians to think of mental disorders as clustering in families, official psychiatry would prompt them to study, debate and ultimately seek out implications tied to the generative processes being proposed as the bases of the clusters—processes that are proposed to either evoke or sustain the conditions, that rest sometimes on cerebral changes and sometimes on life circumstances and that, if confirmed, will inform rational treatment and prevention.

Such a modest rearrangement and thematic reorganization of disorders would persuade psychiatrists to think much more "contextually" about their patients and their patients' disorders. This would prompt them to reject symptom checklists as diagnostic instruments in favor of more thorough assessments of their patients that specifically consider their social environments and developmental trajectories, wherein lie the generative sources and differentiating information about many of their disorders.

An official classificatory system should do more than name and list disorders. It should organize them in ways that propose the modes of study most likely to explain them. Such revision of DSM would bring direction back into psychiatric thought, practice and research—indeed it would impel psychiatrists in such a direction by its implications. Psychiatrists should not be satisfied—especially after 30 years—with a process that runs on the hope that diagnostic consistency alone will eventually translate into explanations. This approach has failed for more than a generation to deliver discoveries that can amend it.

Psychiatry may not have what it takes to form a unified "theory" of mental disorders, but it has concepts with enough credibility to indicate that certain disorders differ in their fundamental nature and that these differences are sufficient to influence treatment decisions and to suggest the most apt ways of study. If DSM-V turns out to be nothing more than a tinkered-with version of DSM-IV, many will view it as a failure of nerve.

Perhaps more than anything else, such a simple classificatory rearrangement—by encouraging reflection on the part of psychiatrists and patients alike about the generation of mental disorders (and mental states that are not disorders)—would bring back to psychiatry what we see DSM-III/IV having taken away: a body of assumptions and debatable issues to be ever thought over, contemplated for their implications and sifted through in discerning approaches to better practices and critical research. Regaining this essentially contemplative aspect of the discipline will bring renewed investigative energy to psychiatry and, far from unimportantly, make it fascinating again.

8

Using Deep Brain Stimulation on the Mind

Handle with Care

By Mahlon R. DeLong, M.D.

Mahlon R. DeLong, M.D., is the William Timmie Professor of Neurology at the Emory University School of Medicine and a member of the Dana Alliance for Brain Initiatives. His research has contributed significantly to the revival of and development of new and more effective surgical approaches for the treatment of movement disorders. Dr. DeLong has received awards for his research and clinical contributions. He is an elected member of the Institute of Medicine and the American Academy of Arts and Sciences. Health America recognizes him as one of the top doctors in neurology for the treatment of movement disorders.

Deep brain stimulation has worked for many patients with Parkinson's disease and other movement disorders that have not responded to other treatments. However, its use as a therapy for psychiatric disorders, while promising, is not yet proven. Mahlon DeLong, a pioneer in the use of deep brain stimulation, explains the technique and why its use for depression, obsessive-compulsive disorder and other psychiatric problems requires extra caution.

NEUROSURGICAL TREATMENTS for movement disorders such as tremor and Parkinson's disease have undergone a renaissance in recent decades, first with ablation (the surgical removal of targeted brain tissue) and more recently with minimally invasive deep brain stimulation (DBS). Ablative surgery laid the groundwork for the introduction of experimental DBS in the treatment of several motor and psychiatric disorders, such as severe depression and obsessive-compulsive disorder, by pinpointing the specific networks of neurons involved. DBS modulates abnormal activity in these circuits, and preliminary studies of DBS in patients who fail to respond to conventional therapies for these disorders have shown promise.

Although more than 56,000 patients have undergone DBS surgery, mainly to treat movement disorders, DBS has received Food and Drug Administration (FDA) approval only for use in treating Parkinson's and essential tremor. The procedure's early successes have led desperate patients and families, as well as companies that manufacture the hardware, to seek more widespread approval. Researchers must balance this pressure with the need to perform well-controlled clinical studies to determine whether DBS is a safe and effective treatment for various psychiatric illnesses. Maintaining this balance will be essential for making research progress while ensuring that treatment decisions can be based on sound scientific evidence.

How does DBS modulate activity in brain circuits? Surgeons implant electrodes that can be stimulated electrically into specific brain regions. An implanted, but externally programmable, pulse generator similar to a cardiac pacemaker then delivers continuous, high-frequency electrical stimulation to the brain tissue through the electrodes. DBS counteracts the motor abnormalities of Parkinson's disease—slowness of movement, resting tremor and muscular rigidity—and other movement disorders such as essential tremor and dystonia. It offers significant advantages over earlier surgical techniques because it is reversible and adjustable. Because DBS influences the activity of brain circuits, it is now usually referred to as "neuromodulation," a change that emphasizes its more restorative and less destructive effects on the brain. Scientists are also considering DBS as a potential treatment for other neurological problems, including epilepsy, pain, cluster headaches, obesity and even cognitive impairment.

Initially, scientists thought that DBS simply silenced the targeted brain circuitry, mimicking the effects of lesioning. Instead, we have learned that DBS stimulates or activates targeted brain regions and acts to normalize neural network activity in those regions. The effects of DBS are significantly more complex than we initially thought, and the treatment may work differently depending on the brain regions and networks involved. When intervening in circuits in the brain's basal ganglia, for instance, DBS has been shown to effectively treat Parkinson's patients' slow movement, resting tremor and rigid muscles. It also has helped individuals with essential tremor and dystonia, a disorder characterized by involuntary twisting movements and abnormal postures. Patients with these three different movement disorders respond to stimulation of the same DBS targets of the motor circuit, which suggests that the effects of this treatment are not disease- or symptom-specific, but rather circuit-specific.

The effective use of DBS in the basal ganglia to treat these movement disorders has been expanded to the experimental treatment of several psychiatric disorders. The basal ganglia are components of a family of larger brain circuits that are involved in different aspects of behavior. These networks are involved not only in movement, but also in executive functions (such as decision-making and planning), mood regulation, reward and motivation, the primary targets of experimental DBS treatment for psychiatric disorders.

Researchers currently are studying DBS treatment primarily for three psychiatric disorders: severe depression; obsessive-compulsive disorder (OCD), which is characterized by intrusive thoughts and compulsive behaviors and rituals such as hand washing and counting; and Tourette's syndrome, which is characterized by a combination of involuntary vocal and motor tics and associated psychiatric disturbances including depression, hyperactivity and obsessive and compulsive behaviors. In February 2009 the FDA granted a Humanitarian Device Exemption—a form of limited approval for treating conditions affecting fewer than 4,000 people per year in the United States when no other effective treatment is available—for treating patients with severe OCD (the first such approval for a psychiatric disorder in the U.S.) and patients with dystonia, a sometimes profoundly disabling movement disorder. This exemption allows clinical researchers to use DBS in patients with these conditions once the investigators have received approval from their hospitals' or universities' institutional review boards.

DBS treatment for psychiatric disorders introduces additional clinical and ethical challenges because the brain targets are circuits involved not simply with movement but with behavior, mood, motivation and reward—the core of our being. Because we are sure neither how DBS restores brain function and reduces clinical abnormalities, nor what its long-term effects are, researchers and surgeons must proceed with caution.

Success Comes with Caveats

Our experience with DBS as a treatment for movement disorders is instructive, because even though DBS has proved successful on the whole, we have learned that it is a complicated procedure with significant risks and side effects that can be both physical and psychiatric. Although most problems stemming from DBS surgery are temporary, 1 to 2 percent of patients experience major complications.

DBS for Parkinson's, for instance, can cause involuntary movements, impaired cognition and word finding and worsening of speech, gait and balance, but changing the level of stimulation can often remedy such problems. Seizures and failures of the DBS hardware, though never widespread, have become even rarer as techniques and equipment have improved. According to some studies, however, 5 to 10 percent of patients experience infection under the skin where the electrodes and the stimulation device are implanted.[1]

By the time patients with movement disorders are referred for surgical evaluation, many have already decided to go ahead with DBS. However, these patients need to talk with their doctors about their expectations, their prior medical and psychiatric problems and the known risks and potential benefits of DBS. In the end, some patients fail to benefit from the surgery. These failures most often result from improper selection of patients, poor electrode placement and programming errors. Early in the development of DBS for movement disorders, some neurosurgeons operated on patients without the benefit of careful screening by a movement disorder specialist and adequate supporting practitioners. Such problems have become less frequent as patients have found established movement disorder centers that conduct surgical procedures and have had their devices reprogrammed or, when necessary, have undergone a second surgery to replace an off-target electrode.

Experience with performing DBS on patients with movement disorders has demonstrated that it is crucial for patients to seek a full explanation of the procedure's potential benefits and risks. It is also important that patients have an experienced team: a neurologist who specializes in movement disorders, a well-trained and experienced neurosurgeon, a psychiatrist, a neurophysiologist, a device programmer and experienced physical and speech therapists. This team bears the major responsibility for screening and selecting suitable candidates, educating patients and their families about the surgery, setting realistic expectations and fully discussing risks and possible side effects. Not all patients with Parkinson's and other movement disorders are suitable candidates, and this is equally true of people with psychiatric disorders. Careful screening and selection of patients are fundamental requirements for success.

Implications for Psychiatric Disorders

In *rare* instances, DBS itself can exacerbate or possibly cause depression and abnormal behaviors in Parkinson's disease patients, typically when the electrode is placed in a location that allows the stimulation to spread beyond the intended regions. These instances have foreshadowed the heightened complexities of experimental testing of DBS in patients with psychiatric disorders. A recent large, multicenter retrospective study of patients with Parkinson's disease who received DBS has revealed an increase in the risk of suicide, attempted suicide and postoperative depression. This study highlights the importance, both before and after DBS surgery, of carefully screening Parkinson's disease patients for psychiatric risk factors, especially depression, and aggressively treating any disorder found before and after DBS treatment commences. Patients with movement disorders who suffer from depression may not feel they have benefited as expected from DBS, even though they clearly have according to objective measurements.

Although experience with treating movement disorders has provided guidance for the experimental use of DBS in treating some psychiatric conditions, and although the two types of disorders have some common problems, there are still open questions about how to proceed in this new era of neuromodulation. How can scientists, medical professionals and the public balance the desire for scientific progress and the need to protect patients from undue risks? In spite of intense pressure for widespread use of DBS, pilot studies and subsequent, well-planned, large clinical trials must take place first, and these can only happen in centers with fully staffed and dedicated teams. Patients and researchers alike must be committed to long-term care, with frequent follow-up visits for programming and medication adjustments. Only after careful research and dissemination of trustworthy findings can an informed public, along with surgeons and physicians who work with the best advice available from colleagues, ethicists, professional societies and lay organizations, help determine and then ensure the proper use of DBS for psychiatric disorders.

To better understand the potential pitfalls of widespread use of surgical approaches without adequate clinical studies, we need only to look at the long, controversial and sometimes sordid history of modulating mood and behavior with neurosurgery. In particular we may consider frontal lobotomy, a treatment option in the 1950s for patients with severe psychotic disorders such as schizophrenia. Neurosurgeons thought that severing the excitatory fibers in the lower brain of an agitated patient would calm him, but the surgery often left a patient passive and unmotivated—essentially a "vegetable." Today we consider this procedure barbaric, but at that time—before the discovery of antipsychotic drugs—a lobotomy was the only alternative to physical restraints or seclusion in a mental hospital. The use of this procedure after World War II was part of a crisis in treating chronically mentally ill people. The situation was so severe that more than half the public hospital beds in the U.S. were used to treat patients with mental illnesses. With the introduction of psychotropic medications such as chlorpromazine (Thorazine) in the 1950s, the use of lobotomies declined precipitously. Eventually, large numbers of patents left state hospitals and received care in community health centers.

Lobotomies are but one example of early surgical explorations, for both psychiatric and movement disorders, that came about because surgeons were willing to explore potential brain targets and patients were desperate enough to take significant risks. Several factors led to a near abandonment of such surgical procedures worldwide: the introduction of levodopa for the treatment of Parkinson's disease, the discovery of the first antipsychotic and antidepressant medications and the public and professional reaction in the 1960s against the excesses of "psychosurgery." More medically and scientifically informed surgical lesioning procedures appeared again only in the early 1990s, due to growing appreciation of the enormous burden of neurological disorders, the frequent failure of medical therapies to ease that burden and the side effects of many of the available drugs.

Evidence for Cautious Optimism

Despite the lessons of the past and the unknowns associated with DBS, the procedure appears to be mind restoring rather than mind altering. Preliminary study results suggest that DBS relieves the painful weight of depression and the domination of obsessive thoughts, compulsions and urges, and thus restores a more normal level of functioning. There is no evidence that long-term electrical stimulation causes cell damage or permanent changes in the networks stimulated. Although DBS induces long-term changes in circuit function, these alterations seem to be reversible, as demonstrated by the fact that disorders typically re-emerge when a DBS battery fails or a lead breaks.

In some cases—especially if the patient is a minor—pre-existing mental illness may compromise a patient's ability to understand both the key aspects of DBS and the risks involved. Thus, the all-important treatment team also should include an ethicist, either as a member or an advisor. In addition, as has been the case for movement disorders, clinical trials must have strict guidelines that exclude many patients. Patients and their families may be desperate, but without a careful selection process, we will not be able to identify the disease characteristics that indicate which types of patients are most likely to benefit from DBS.

In general, patients seeking experimental DBS surgery for major depression must have had the disorder for at least a year and must first have tried multiple treatments, including electroconvulsive (shock) therapy. Their chronic depression must also be relatively uncomplicated; patients with significant paranoia, anxiety or panic attacks would be excluded, for example. The careful screening process includes formal interviews and neuropsychological testing, as well as a visit with an independent psychiatrist, whose discussions with the referring psychiatrist and the patient's family help build consensus and understanding and help ensure stable long-term care.

Although preliminary results of DBS for treating psychiatric disorders are encouraging, scientists understand the workings of the corresponding brain circuits far less than they understand the circuits involved in movement disorders. For patients with most psychiatric disorders, the chances of improvement via DBS are currently less than 50/50. There is considerable uncertainty. Whereas movement disorders, with the exception of dystonia, show a relatively rapid response to DBS, depression, Tourette's syndrome and OCD are generally slow to respond; effects can take weeks or months to appear, and this delay can be frustrating and worrisome.

Even success can pose problems. As is true for patients with movement disorders, people with psychiatric disorders who, thanks to DBS, become more functional and independent after a long period of disability may experience disruptions in their relationships with spouses and other caregivers. Many people with depression and OCD have never known what it is like not to be severely afflicted, and their improvements are painfully slow and uneven. Furthermore, the adverse effects of undetected circuit stimulation failure in depressed patients are potentially far more severe than those involved in movement disorders. These effects can include severe mood changes and suicide attempts.

The two research groups studying different DBS targets in the brain for treating depression receive support from competing device manufacturers. It will be important for the groups to compare their approaches in an unbiased and rigorous way to determine which achieves the better outcomes and the less troublesome side effects. Each of the two circuit targets may prove to be more or less appropriate, depending upon the type of depression and the patient's symptoms. Future studies should compare not only the benefits and side effects of stimulating each target, but also the nature and duration of disease remissions and patients' long-term outcomes.

For patients with Tourette's syndrome, the criteria for surgery are more complex because the disorder's severity usually decreases after adolescence. Because symptoms often diminish, surgery is not recommended for patients age 25 or younger, except in the most severe cases. The risks of DBS surgery for younger patients often outweigh the potential benefits.[2] Although most clinical trials have focused on controlling patients' tics, researchers recognize that the associated psychiatric disturbances often add considerably to the disorder's burden and, if they are extreme, may justify a second look at surgery.

Researchers have seen promising preliminary results from early DBS clinical trials involving patients with OCD, as well as patients with Tourette's syndrome who have obsessive-compulsive behaviors, but these patients illustrate another potential complication: the obsessive symptoms themselves. For example, patients given the option to vary their DBS parameters themselves may do so compulsively. Others may damage the device by repeatedly manipulating or picking at the skin that covers the implanted battery or the connecting wires.

Before researchers can recommend the use of DBS for routine treatment with Tourette's syndrome and OCD patients, they must carefully explore the effects of stimulating different targets in the brain and develop patient selection guidelines and postsurgical management plans. It is encouraging that investigators at different institutions have collaborated to carry out clinical trials and to publish statements regarding patient selection, inclusion and exclusion criteria, preoperative and post-operative evaluations and selection of brain targets for the treatment of Tourette's syndrome and OCD. These proactive efforts reduce the risk that the application of DBS surgery for psychiatric disorders will outpace the science that supports it.

Next Steps

As was the case for movement disorders, pilot studies in a small number of patients whose psychiatric disorders have been well characterized are critical to explore the different potential brain targets and to assess the initial safety and efficacy of DBS. Evidence from such studies suggests that DBS does benefit patients with severe mental illnesses. This preliminary evidence justifies undertaking larger-scale clinical trials in which patients are randomly assigned to receive DBS or a placebo (wherein DBS electrodes are implanted but not fully activated) to establish safety and efficacy in implementing the surgery. By administering subthreshold stimulation to the placebo group and a full therapeutic level of stimulation to the experimental group, clinical researchers can identify the treatment's real effects. Such larger studies also can reveal side effects that are less common but significant.

We do not know exactly how the future of DBS and the new field of neuromodulation will evolve, but the genie is out of the bottle. Increasing numbers of patients will undergo treatment with DBS for some of the most distressing and disabling disorders known to mankind. Early successes and rapid progress generate both hope and concern.

First, hope: It may soon be possible for patients with the most disabling and resistant forms of depression, OCD and Tourette's syndrome to find relief. Moreover, progress may spur experimental use of DBS for other psychiatric disorders. Now, concern: Scientists must avoid repeating the mistakes of the past and withstand the pressures to proceed too quickly into uncharted waters. Guidelines are essential to protect patients, and they must include rules for comprehensive evaluation and full discussion of risks, benefits and alternative approaches to ensure that DBS is the appropriate treatment choice. Careful oversight and monitoring by institutional review boards is also essential. Ultimately, physicians and psychiatrists must assure the public that they will do away with the negative associations of *psychosurgery* by bringing the modern field of *neuromodulation* to its full potential.

9

Neuroimaging

Separating the Promise from the Pipe Dreams

By Russell A. Poldrack, Ph.D.

Russell A. Poldrack, Ph.D., is a professor in the department of psychology and the department of psychiatry and biobehavioral sciences at the University of California, Los Angeles. His research entails using neuroimaging to examine brain systems that underlie our abilities to learn, to make decisions and to control our behavior. He is one of the leading theorists on how to interpret these neuroimaging results. In addition to studying brain function in healthy individuals, Dr. Poldrack examines the brain in the context of neuropsychiatric disorders such as schizophrenia, attention-deficit/hyperactivity disorder (ADHD), Tourette's syndrome and drug abuse.

Colorful brain images may tempt researchers to make claims that outpace solid scientific data—and may tempt the public to believe those claims. In particular, although brain imaging has provided solid evidence of alterations in brain structures and functions associated with many psychiatric disorders, it can be used neither to diagnose such disorders nor to determine exactly how treatments work—at least not yet. Keeping some key ideas in mind can help us evaluate the next report of a brain-imaging "breakthrough."

ON ANY GIVEN DAY you are likely to see a news report mentioning brain imaging. As I write this, a quick search of recent news stories yields the following headlines:

- "Justice May Be Hard-wired into the Human Brain" (New Scientist)
- "Brain Area Blamed for Stress Disorders" (RedOrbit.com)
- "School Bullies—Is the Amygdala to Blame?" (BrainBlogger.com)

Neuroimaging research clearly has captured the imagination of both the public and science writers. Given how far brain imaging has come in the last two decades, this is understandable. Functional magnetic resonance imaging (fMRI) has revolutionized our ability to image brain activity safely, and its broad accessibility has allowed researchers around the world to ask fascinating new questions about the mind and brain. At the same time, it is all too easy to leave the limitations and caveats of these methods out of the picture. This results in a common perception that overrates the power of brain imaging to explain everything from love and beauty to financial decision-making.

One of the main limitations of neuroimaging is that conclusions based on studies of groups of people might not apply to individuals. This limitation becomes especially important when the use of neuroimaging moves beyond the realm of scientific generalities to the domain of individual health care decisions. The apparent power of neuroimaging can be overwhelming to a parent searching for an explanation for her child's disruptive behavior, or to a child seeking answers about his parent's memory loss. However, many proposed applications—particularly those relating to the diagnosis or treatment of psychiatric disorders—simply are not supported by evidence from research.

What Neuroimaging Can and Cannot Tell Us

Functional magnetic resonance imaging is a powerful tool. It measures blood flow in the brain, which increases when the neurons become active, thereby indirectly measuring their activity. For example, one might use fMRI to image brain activity while research participants are engaged in a mental task such as rehearsing a seven-digit number, and then compare the results with scans taken while the participants rest quietly. Using statistical methods to compare these images, we can determine where in the brain the difference is strong enough to be considered important. The colorful "activation maps" we see in news stories highlight these areas of the brain. We also can compare the brain activity of two different groups of people—for example, healthy individuals and individuals with schizophrenia—to determine which brain areas differ.

Neuroimaging research has provided fundamental insights into human brain function and mental disorders, and nearly every area of psychology and psychiatry has changed as a result. For example, drug-abuse researchers have used neuroimaging to isolate a set of brain systems in which an abnormal response to rewards is associated with drug dependence. In principle, this research should lead to improved drug-dependence treatments that specifically target the isolated brain systems. It could also allow doctors to predict whether specific treatments will be effective, as researchers have created a biological index (or biomarker) for the function of these brain systems.

Neuroimaging research also has provided insights into brain development. Studies using fMRI have linked adolescence to increased activity in reward systems, while the development of prefrontal cortex functions, such as judgment and impulse control, lag behind. This difference helps explain the prevalence of impulsive and reward-driven behavior in adolescents. (See Chapter 5, "The Teen Brain: Primed to Learn, Primed to Take Risks.")

Despite this progress, neuroimaging leaves several questions unanswered. First, it is impossible to say whether increased or decreased activity in a particular brain region is "better" or "abnormal." For example, some studies found that schizophrenia patients had increased activity in specific brain areas (such as the prefrontal cortex) compared with healthy individuals, while other studies found that patients had decreased activity in the same brain areas. These conflicting findings suggest that group differences may be specific to the tasks on which the groups were tested, and they cannot be interpreted broadly as reflecting "better" or "worse" brain function.

Second, we cannot assume that individual brain areas are uniquely responsible for specific mental functions, and thus that activation of those regions tells us what a person is thinking. A prime example of overreaching is the application of neuroimaging to an emerging field known as neuromarketing. On its Web site, the neuromarketing firm FKF Applied Research claims that its researchers have used fMRI to map a set of specific mental processes onto specific brain regions, and the firm can use those maps to determine how individuals respond to stimuli such as advertisements:

> Over the past decade, fMRI has allowed us to "map" several key regions of the brain with a high degree of specificity. ... A key part of that data is how the brain reacts in 9 well known and well mapped areas, such as the Ventral Striatum (reward), Orbitofrontal Prefrontal Cortex (wanting), Medial Prefrontal Cortex (feeling connected), Anterior Cingulate Cortex (conflict) and the Amygdala (threat/challenge). Using this mapped data, as well as data from other parts of the brain, we have developed a set of norms that help us understand what is happening inside a subject's brain when they are exposed to a particular type of stimuli.[1]

This claim relies on a simple logical error. The fact that the amygdala, for example, responds to threat does not mean that activity in this area signifies that a person is feeling threatened. That would be true only if threat were the *only* thing that activates the amygdala, and we know this is not the case.

Although this article focuses mainly on neuroimaging in the diagnostic context, other areas in which the implications of neuroimaging results are being cited range from love and happiness to creativity to decision-making. Such applications often rely upon exactly the same kind of flawed logic. For example, an article in *Esquire* magazine ("Do I Love My Wife? An Investigative Report," May 18, 2009) describes the author's participation in an fMRI study that examined how his brain responded to photos of his wife in order to "scientifically assess [his] love." The researchers claimed to be able to determine the degree to which the writer felt lust, romance and attachment for his wife by looking at which regions of his brain were active while he viewed the photos. However, the regions these researchers identified are the same areas that other researchers have associated with different functions; the claims about the writer's love for his wife based on these scans are almost certainly overblown.

Diagnosis and Prediction Using Neuroimaging

Brain imaging has revolutionized psychiatry by providing compelling evidence for the biological basis of psychiatric disorders. A large and increasing body of research has shown that individuals diagnosed with psychiatric disorders show specific differences in brain structure and function when compared with healthy individuals. These findings lead us to imagine that the use of brain scans to diagnose psychiatric disorders is just around the corner. A closer look, however, demonstrates the problematic nature of this logical leap.

As an example, imagine a study that compares brain activity in response to monetary rewards in healthy individuals with that of drug abusers. Researchers analyze the data to determine whether there is a statistically significant difference in activity between the groups. In this context, statistical significance means that such a difference in observed activity would be very unlikely (less than a 5 percent chance) if there truly were no difference between the groups.

Let's say that activity in a particular brain region was greater in the drug abusers than in the healthy individuals. We cannot accurately predict who is a drug abuser based on the fMRI data alone, because some healthy individuals' levels could be above the average for drug abusers, and some drug abusers' levels could be below the average for healthy individuals. Without doing more specialized statistical analyses, we cannot reliably predict drug abuse from observations of brain activation.

A new set of techniques, however, offers the potential to classify individuals more reliably as healthy or not based on their brain activation. These come from a new field of statistics called machine learning. In the past few years, scientists have applied machine-learning analyses to fMRI data, and these analyses have shown some ability to accurately diagnose individuals with psychiatric disorders. However, none of them has been subjected to the rigorous tests that would be necessary to supplant standard diagnostic approaches.

Neuroimaging research also could completely change how we think about psychiatric disorders by rendering obsolete the idea that using discrete diagnostic categories such as schizophrenia or attention-deficit/hyperactivity disorder (ADHD) provides the best way to understand the underlying disorders. Today, these diagnoses are based on formal criteria, outlined in the American Psychiatric Association's *Diagnostic and Statistical Manual of Mental Disorders*, that specify symptoms for each disorder. But these criteria have no basis in neuroscience. In fact, the psychiatric community has become increasingly concerned that traditional diagnostic categories actually obscure the underlying brain systems and genes that lead to mental health problems. In addition, a growing body of evidence indicates that many psychiatric problems lie on a continuum rather than being discrete disorders, in the same way that hypertension reflects the extreme end of a continuum of blood pressure measurements. Neuroimaging provides us the means to go beyond diagnostic categories to better understand how brain activity relates to psychological dysfunction. However, using neuroimaging to "diagnose" classical psychiatric disorders could obscure, rather than illuminate, the true problems.

Another potentially important application for neuroimaging is to determine the best treatment for individuals with mental health disorders.

For example, some children with ADHD do not respond to the stimulant medications that doctors usually prescribe for this disorder. It would be useful to be able to predict which children will benefit from which kinds of medication. Evidence remains insufficient to support this kind of prediction.

Nevertheless, the use of brain imaging has been championed to optimize treatment of psychiatric disorders. For example, Daniel Amen, M.D., has clinics that charge several thousand dollars to perform single photon emission computed tomography (SPECT) scans—which involve the injection of a small amount of a radioactive tracer—on individuals with one or more of a wide range of disorders, with the promise of "tailor[ing] treatment to your brain." For example, Amen says on his Web site that children with ADHD who have a "ring of fire" pattern of brain activity respond better to anticonvulsant or antipsychotic medications than to stimulant medications. Amen's own research, however, reports accuracy of less than 80 percent in predicting treatment outcomes. And even this number is probably an overestimate, as the study in question—which involved 157 children overall—excluded 120 patients who did not show activation in a particular brain region.[2]

Given the scant evidence, it seems unlikely that the potential benefit on treatment outcomes justifies exposing children to even the small amounts of radiation from SPECT scans. The American Psychiatric Association affirms this opinion in a 2005 report: "[T]he available evidence does not support the use of brain imaging for clinical diagnosis or treatment of psychiatric disorders in children and adolescents."[3]

Someday neuroimaging will provide a better means to tailor treatment for psychiatric disorders. Until proper clinical trials show its effectiveness, however, we must regard this powerful technology with healthy skepticism.

Evaluating Neuroimaging Research

Brain imaging will only continue to penetrate our 21st-century lives. Because it reminds us that mental disorders are biologically based, this development is positive. However, it is important for readers and consumers to understand how scientists study and apply neuroimaging.

First, publication in a peer-reviewed journal is necessary for any research finding to be taken seriously. Peer review is a cornerstone of modern science, the gatekeeper that helps ensure that any published research finding meets accepted standards for scientific methodology and reasoning. The peer review process is not perfect, however, and inappropriate claims and inaccurate research sometimes fall through the cracks.

Papers in the most prominent journals are sometimes the biggest offenders when it comes to over-interpretation. Because these journals tend to publish findings of substantial general interest, authors often present strong—and potentially inaccurate—interpretations of their results. "Special issues" of peer-reviewed journals also are prone to publishing flawed papers. In such issues the usual peer-review process is often weaker, as the guest editor may be reluctant to reject solicited submissions.

Second, it is important that any research finding be replicated by multiple independent groups of researchers before it is used as the basis for decisions about medical treatment. This is especially true for studies using small sample sizes, which is often the case in neuroimaging research due to its substantial expense. Defining a sample size as "small" is difficult because it depends on the scale of the effect that is being measured, but for standard neuroimaging studies, we should treat with skepticism any finding based on fewer than 20 participants.

Third, researchers must show how differences in brain activation—either differences between participant groups or differences resulting from some intervention—relate to behavior or cognitive function. Such brain activation differences are not, on their own, compelling. Meaningful examination of group differences requires that groups be tightly controlled so that the only difference is the one in question. This type of control can be very difficult in the context of psychiatric disorders. For example, if two groups are tested and one of the groups underperforms, this difference in performance can result in differences in brain activity that do not relate directly to a disorder. Likewise, when individuals are trained on a task and they become better at it, changes in brain activity could simply reflect the fact that they are performing better, rather than reflecting a fundamental change in how their brains are processing information.

Fourth, brain areas do not correspond uniquely with mental functions, as news reports about neuroimaging research often imply. While the activity of specific areas certainly is crucial to such functions as memory or fear, it does not follow that these regions are "memory areas" or "fear areas." These areas are probably involved in other functions as well; conversely, other brain areas are probably involved in memory and fear. Individual brain regions never work independently. The brain is a complex, dynamic system, and the activity of any area can only be interpreted in the context of other active areas.

Finally, we must be aware—and wary—of the commercial interests of the scientists publishing neuroimaging research. For example, many studies have investigated the potential role of fMRI in lie detection. Two of the groups involved in this research have licensed their technology to companies that are developing a commercial product. Before we can take any such results seriously, independent groups that are free of any conflict of interest must replicate the research.

Neuroimaging has the potential to revolutionize how we view our own minds and brains and how we understand the brain dysfunction that results in psychiatric disorders. However, we should keep in mind that these amazing pictures of brain activity are much more complicated than they initially appear, and interpreting them is never simple. These methods hold the promise of providing guidance for individual medical treatment, but that promise has yet to become a reality.

10

Why So Many Seniors Get Swindled

Brain Anomalies and Poor Decision-making in Older Adults

By Natalie L. Denburg, Ph.D., with Lyndsay Harshman, B.S.

Natalie L. Denburg, Ph.D., is an assistant professor of neurology and neuroscience at the University of Iowa Carver College of Medicine. Her research interests involve the neural basis of decision-making abilities in older adults; consumer, medical and financial decision-making; neuroepidemiology; social and affective neuroscience; and cancer survivorship.

Lyndsay Harshman, B.S., is a third-year medical student at the University of Iowa Carver College of Medicine. She was awarded a Doris Duke Clinical Research Fellowship for the 2008–2009 academic year.

"You won't believe what happened!" an elderly man said to his son. The father had received a telephone call informing him that he had won a large prize. To collect it, all he had to do was wire $4,000 to a Guatemalan bank. He had wired the money before he called his son.

A few days later and several states away, a "nice man" knocked on an older woman's door and offered to repair her storm-ravaged roof for $6,000. Without verifying his identity, she handed him a check.

Although decision-making abilities often decline as we get older, Natalie Denburg argues that we should not consider such deficits to be a part of "normal aging." Nor are they likely signs of a dementia such as Alzheimer's disease. Denburg's research suggests that some older adults experience flawed emotional responses that stem from abnormalities that develop in the brain's prefrontal cortex. Further research could help identify and protect people who are especially vulnerable to being swindled.

DECEPTIVE AND FRAUDULENT advertisers, telemarketers and door-to-door salespeople are notorious for targeting older adults. Stories of vulnerable people losing their money, and their sense of dignity, shock and sadden us.

We do not wholly understand why elderly people are vulnerable to such schemes. Although possible explanations range from loneliness and gullibility to memory impairment and dementia, these characteristics do not accurately describe many of the victims. For example, some older adults have provided cogent congressional testimony in which their ability to describe their experiences with multiple scams does not give any evidence of memory impairment or generalized dementia.

Instead, studies using brain imaging suggest that a subset of older adults who have no diagnosable neurological or psychiatric disease may experience disproportionate, age-related decline in specific neural systems crucial for complex decision-making. New functional neuroimaging findings, along with results from behavioral, psychophysiological and structural imaging studies of the brain, indicate that these seniors may be losing their ability to make complex choices that require effective emotional processing to analyze short-term and long-term considerations. Older adults in this category fall prey to the promise of an immediate reward or a simple solution to a complicated problem. They fail to detect the longer-range adverse consequences of their actions. Finally, they may assume long-term benefits in situations where there are none. We see these characteristics as direct consequences of neurological dysfunction in systems that are critical for bringing emotion-related signals to bear on decision-making.

Could an older adult who makes poor choices be in an early stage of Alzheimer's disease or at greater risk for developing the disease? Although we cannot definitively answer the second part of this question, our studies indicate that isolated impaired decision-making among older adults is a discrete phenomenon. It is distinct from Alzheimer's in terms of the brain regions affected, the course and progression of the syndrome, and the likely brain abnormalities involved. Furthermore, its clinical symptoms appear to be far more subtle than those of a dementia.

While researchers must pinpoint the causes of decline in decision-making ability in more detail, we believe that health care professionals and family members can help to identify potential victims and to take preventive measures now.

"Healthy" but Vulnerable

There are many theories on how a "healthy" brain ages. Some of these ideas contradict conventional wisdom, which holds that aging is synonymous with memory loss. Although the human memory does tend to deteriorate modestly with age, many older people experience far more dramatic declines in cognitive abilities that are not related to memory, such as concentration, problem solving and decision-making. Unlike the ability to remember, which scientists have linked to the medial temporal region of the brain, these other abilities are closely associated with the frontal lobes.

A recent theory called the frontal lobe hypothesis[1] proposes that some older people have disproportionate, age-related changes of frontal lobe structures and the cognitive abilities associated with those structures. Several sources of evidence, including neuropsychological, neuroanatomical and functional neuroimaging studies, support this theory.

Following up on the frontal lobe hypothesis, our team of scientists at the University of Iowa has proposed that some older adults are vulnerable to fraud because they experience disproportionate changes in certain areas of the anterior portion of the frontal lobes. These areas include the ventromedial prefrontal cortex, a vast expanse of highly evolved brain tissue. Along with scientists from other laboratories, we have demonstrated that damage to this subregion—due to stroke, tumors or other injuries—can cause dramatic changes in personality and higher-order abilities: reasoning, judgment, decision-making and emotional processing.

Our observations have led to a theory of how the prefrontal region influences certain aspects of higher-level cognition: the somatic marker hypothesis.[2] The term *somatic* refers to body- and brain-related signals, which we experience as emotions and feelings. According to the somatic marker hypothesis, we make choices that are in our best interest only after we effectively weigh potential short-term and long-term outcomes. A key idea of this hypothesis is that when a decision's outcomes are ambiguous or uncertain, a person's emotions and feelings are essential to making a decision.

The ventromedial prefrontal cortex is critical in triggering various bodily changes (somatic states) in response to stimuli such as cues for reward (a positive outcome) or punishment (a negative outcome). Another area that participates in this process is the insular cortex, where such bodily changes are represented in the brain. As we make decisions in uncertain conditions, our assessment of immediate and future potential consequences may trigger numerous conflicting responses. For example, a highly favorable potential consequence may trigger excitement and elation, while an unfavorable consequence may trigger pain and dread. The result, however, is the emergence of an overall positive or negative signal—basically, a message of either "go" or "stop."

We propose that the brain may trigger numerous and conflicting signals simultaneously, but sooner or later, stronger signals trump weaker ones.[3] In this way, emotional processes are critical for making good long-term decisions. People deprived of appropriate emotional signals—because of damage to the ventromedial prefrontal cortex or the insular cortices, for example—may fail to perceive potential adverse long-term consequences.[2] Thus, we believe that too little emotion can undermine effective decision-making.

To test this neural theory of faulty decision-making, we have used an extensive battery of neuropsychological tasks. For example, the Iowa Gambling Task provides a close analogue to real-life decision-making by setting up situations with rewards, punishments, and unpredictability. The task taxes decision-making functions that are mediated by the ventro-medial prefrontal cortex[4] because it requires the participant to forgo immediate and alluring rewards in favor of rewards that provide greater long-term benefits. Using skin conductance—an index of physiological arousal collected via electrodes placed on the palms of the hands—we can measure a person's emotional arousal during the task.

We have found that one-quarter to one-third of seemingly healthy older adults perform the task the way neurological patients with ventro-medial prefrontal cortex injury do: Both exhibit a preference for choices that lead to high immediate reward but greater long-term punishment. We call this seemingly healthy subgroup the "impaired" decision-makers. In contrast, the larger subgroup of older adults selects choices that have low immediate reward but higher long-term reward. We call this group the "unimpaired" decision-makers. In other studies, we have found that impaired decision-makers have difficulty discriminating between advantageous and disadvantageous choices, also reminiscent of findings involving neurological patients with ventromedial prefrontal cortex or insular damage.

We gave impaired and unimpaired groups a consumer task in which participants read a booklet of advertisements like those they might find in a typical magazine. The booklet is a mixture of nondeceptive ads and ads that the Federal Trade Commission has deemed deceptive or misleading. We found that the impaired decision-makers showed poorer compre-hension of deceptive ads—and were more likely to say that they would purchase the featured products—compared with unimpaired decision-makers.[5] This study provides direct evidence that the subset of older adults with impaired decision-making abilities (as defined by the Iowa Gambling Task) is particularly vulnerable to fraudulent advertising.

Pinpointing Specific Brain Regions

To complement neuropsychological studies, scientists have used neuroimaging to link certain behaviors to specific parts of the brain . Our team has used imaging to explore the inner workings of the brains of older impaired decision-makers. One technique, magnetic resonance imaging (MRI), enables us to examine the brain's structure in high-resolution. Comparing the MRI results of 20 impaired older adults with those of 20 unimpaired older adults, we found that the impaired group displayed a comparative thinning of an area within the broader region of the ventromedial prefrontal cortex. This region is critical for complex, emotion-related decision-making. This finding is yet another clue about what has gone awry in the brains of elders who make poor decisions.

We recently began using positron emission tomography (PET) to examine the brain's metabolism and cell functioning. PET imaging entails use of a small amount of a radioactive tracer that attaches to glucose and records the brain's utilization of glucose for energy. The resulting picture of differing levels of regional glucose metabolism may indicate abnormal states in the body. (The use of PET has become widespread among doctors diagnosing several neurodegenerative diseases. This is a noteworthy advance because accurately recognizing and differentiating among various types of dementias is a considerable challenge.)

Our ongoing PET imaging study involves 48 older adults, 24 with impaired decision-making and 24 with unimpaired decision-making as measured by the Iowa Gambling Task. To date, our results link intriguingly to our overall framework for explaining why some older adults seem to have impaired decision-making abilities: Several areas of their brains have lower metabolism compared with the brains of older adults who are unimpaired. These brain regions, which include those that are critical for

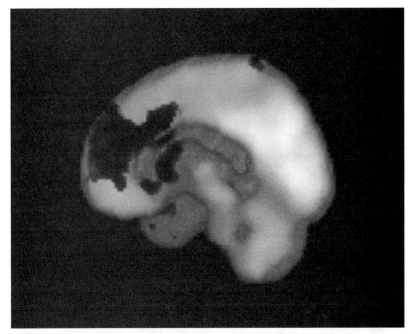

Figure 1 Dark shading from positron emission tomography (PET) scans indicates brain regions in which glucose metabolism was lower on average in impaired elderly decision-makers than in those who were unimpaired. This may reveal an underlying abnormality in the impaired group. (Image courtesy of David Rudrauf, Ph.D., director, Brain Imaging Laboratory, University of Iowa Carver College of Medicine.)

representing emotional states, belong to the brain circuitry involved in the somatic marker hypothesis.

Equally noteworthy, the impaired and unimpaired groups show no consistent metabolic differences in their temporal lobe structures, including medial temporal lobe sectors that are important for memory. These findings, which support our neuropsychological and MRI results, suggest that abnormalities in areas involved in emotions and complex decision-making—rather than areas involved in memory—make some older adults especially susceptible to fraud. To our surprise, the areas involved in working memory, such as the dorsolateral prefrontal cortex, did not appear to be active in the neural representation of decision-making among the older adults we studied. These findings are significant because they suggest that older adults who make poor decisions are not simply "demented," but rather display relatively localized abnormalities in regions of the prefrontal cortex known previously to be important for judgment and making complex decisions.

Possible Interventions

What can we do to identify, protect and help an older adult who is vulnerable to faulty decision-making? In terms of identification, our research findings suggest that someone who is at risk may appear to be of sound mind and body, but family and friends should look for disturbing external signs. Such signs might include the receipt and accumulation of large amounts of mailers with disguised sales pitches, frequent phone and mail-order purchases, large bank withdrawals and dwindling savings. Some affected older adults and their families have set up safety mechanisms such as putting limits on bank withdrawals and personal checks.

Our findings may enhance community efforts to educate consumers and to protect people most at risk of becoming victims of fraud. For example, a public service campaign could show elderly people and their families how to identify predatory marketing practices, respond to potentially fraudulent sales approaches and seek professional help if they fall victim to fraud. Educating older adults also could help them focus on the longer-term consequences of their choices in an effort to decrease their susceptibility to fraud. Testimonials from victims, which can carry strong emotional content, may be particularly effective.

When we and other researchers have more closely identified the neural circuits involved in faulty decision-making, we can develop additional measures to help vulnerable older adults. For example, we might prescribe medications targeting impaired neural systems. Studies involving younger adults link impaired decision-making to abnormalities in brain systems that transmit serotonin and dopamine, chemicals that cells use to communicate.[6] Thus, drugs that target these neurotransmitters may be a potential intervention. Another treatment strategy might involve identifying genetic and environmental risk factors for impaired decision-making.

Unfortunately, we cannot always rely on the patient to report his own problems. People with frontal lobe dysfunction often suffer from impaired awareness and insight (anosognosia); they are unaware of both their own deficits and the ways in which their behavior affects other people. Neurological patients with impaired awareness may deny that they have anything wrong with them, even though their deficits are patently obvious to everyone around them. These patients are particularly liable to place themselves in harm's way, and a significant number of older adults who have fallen victim to financial scams may have such impairment. This makes it more important—and difficult—to detect a person's potential impairment and to design interventions and treatments.

Elderly people with decision-making impairment need more than support from family and friends. They need legal and societal protection from fraud and predatory marketing. We hope our neuroscientific data helps to inform public policy and legislation. As we confirm and extend our research, we anticipate it will prompt strong protections that substantially reduce the number of older adults who are victimized by marketing schemes. We may not be able to eradicate predatory practices completely, but our ongoing research will foster greater understanding of people's potential vulnerabilities and arm us to combat fraud and its consequences.

11

Wired for Hunger

The Brain and Obesity

By Marcelo O. Dietrich, M.D., and Tamas L. Horvath, D.V.M., Ph.D.

Marcelo O. Dietrich, M.D., is a postdoctoral associate in the department of comparative medicine at Yale University. His research focuses on the neurobiology of energy balance. He received his medical degree in 2007 from the Universidade Federal do Rio Grande do Sul (UFRGS) in Porto Alegre, Brazil, and is a Ph.D. student in biochemistry at UFRGS, where he studies the molecular and biochemical pathways involved in brain mitochondrial respiration.

Tamas L. Horvath, D.V.M., Ph.D., is chair and professor of the section of comparative medicine and professor of neurobiology and obstetrics and gynecology at the Yale University School of Medicine. His research focuses on the neuroendocrine regulation of homeostasis, particularly in metabolic disorders such as obesity and diabetes. He received his doctor of veterinary medicine in 1990 from the University of Veterinary Sciences in Budapest, Hungary, and his Ph.D. in 1999 from the Attila Jozsef University of Natural Sciences in Szeged, Hungary.

For most of human history, food was not readily available; storing energy helped ensure survival. Humans thus evolved to eat whenever food is at hand—a tendency that in the modern world may contribute to widespread obesity. Researchers are starting to determine the brain circuitry responsible for this default "eat" message. Marcelo Dietrich and Tamas Horvath tell the story of false starts and measured successes in obesity research. They propose that developing successful obesity therapy may require combining drug therapy with psychological or psychiatric approaches, as well as exercise. In the sidebar, they examine the opposite of obesity: anorexia nervosa.

IN 1994, a scientific finding shot through the research world and then raced far beyond. It inspired wonder and, particularly among discouraged people struggling with severe weight problems, great hope. The report described how a molecule, leptin (from the Greek *leptos,* meaning "thin"), which is present in humans and other animals, powerfully influenced eating in laboratory experiments with obese mice.[1] With good reason, experts speculated freely about defeating obesity—while giving appropriate cautions about the need for more research—and the discussion filled the airwaves and pages of newspapers and magazines for months. But then, just as quickly, leptin was gone. Further research failed to demonstrate a beneficial effect in people, and the "next big thing" was dethroned. People who had hoped leptin would be their salvation were disappointed and frustrated.

But the breakthrough and subsequent experiments with leptin did not leave scientists disappointed—far from it. They had turned an essential corner on the road to understanding the brain and obesity, and researchers have discovered much more in the 15 years since. In this article, we will explain what we now understand about one of the master controllers of this very intricate brain-body relationship. We also offer a surprise ending—of the good news–bad news sort.

Understanding Energy Balance

When thinking about obesity, it is important to remember that consuming food to store energy is fundamental for survival—and nature has done all it can to thwart interference with this mechanism. Much of evolutionary development has been about preserving and synchronizing opportunities to obtain and consume food to maintain a precise balance between food intake and energy expenditure (energy balance).

In the wild, sources of food were few and dispersed widely; early humans had to travel great distances to find food and safe places to rest. Because agriculture developed later on the evolutionary scale (by most estimates, only 10,000 years ago), this migratory behavior required high levels of daily activity. Additionally, because food was scarce, it was essential for us to develop a biological system to store energy. In humans, the largest depot of stored energy is our fat—specifically, the white adipose tissue. Thus, for most of human history, the struggle for life was about seeking, consuming and storing food, and our brains and bodies adapted to thrive in these environmental conditions.

In contrast, humans in modern societies can obtain food on demand and, in many instances, they have low activity levels (low energy expenditure) resulting in what is called *positive energy balance*—we consume more calories than we burn. This chronic, positive energy balance leads to obesity (accumulation of fat tissue) and its associated problems, including diabetes, cardiovascular disorders, cancers and neurodegenerative diseases. A striking development is that in the last few decades, adults have not been alone in their struggles with obesity. Children have become fat at an alarming rate, making childhood obesity a major health issue for most societies.[2] We are seeing an epidemic of positive energy balance.

The Role of the Brain in Maintaining Energy Balance

Several brain nuclei—identifiable groups of neurons functioning together in specific brain areas—regulate the coordination of food intake and energy expenditure. Some of these nuclei are housed within the hypothalamus, a structure that lies just above the brain stem and helps control essential processes such as metabolism, sleep-wake states, body temperature, blood pressure, hunger and thirst and, through connections to other brain circuits, helps regulate brain activities. The hypothalamus was one of the first brain regions to evolve.

One nucleus that helps regulate energy balance is located in the most basal (lowest) part of the hypothalamus and has the form of an arc, giving it its name, the arcuate nucleus (ARC). Researchers have focused their attention on the ARC in recent decades because it has a singular role in energy balance and because it is located in one of the few areas in the brain that is not protected by the blood-brain barrier, the tightly meshed cell structure of blood vessel walls that keeps most blood-borne molecules from entering nearly all brain areas.

Within the ARC is the heart of what scientists believe is the main system for regulating food intake, the melanocortin hormone system. During a meal, various substances are released in the blood, enter the ARC and signal the melanocortin system to end the meal. To sense these substances, the system uses two side-by-side groups of specialized neurons with opposing actions. One neuronal group produces melanocyte-stimulating hormones (MSH) that *suppress appetite*, while the other neuronal group produces molecules that inhibit these hormones' actions and *stimulate appetite*.

MSH hormones signal the brain that no food is needed via a several-step process. The appetite-suppressing neuronal group produces a molecule called proopiomelanocortin (POMC). This molecule ultimately creates large peptides (the building blocks of proteins) that are then broken down by enzymes into small peptides including the appetite-suppressing MSH hormones. These hormones bind to and activate melanocortin receptors located on brain cell surfaces that will signal the rest of the brain that food is not needed.

Two appetite-stimulating molecules, however, oppose the action of these MSH hormones. These two molecules, neuropeptide Y (NPY) and agouti-related protein (AgRP), are produced by neurons adjacent to the POMC cells. These molecules successfully compete against the MSH hormones to bind to the same melanocortin receptors on brain cells and inhibit their activity, so that the rest of the brain no longer receives the signal that food is unnecessary. The interaction between NPY/AgRP (appetite-stimulating) and POMC (appetite-inhibiting) neurons, therefore, regulates food intake.

Moreover, the brain has a redundant system for promoting the urge to eat. Appetite-stimulating neurons also produce the chemical neurotransmitter gamma-aminobutyric acid (GABA), which inhibits the POMC neurons from producing MSH hormones. During periods of negative energy balance, therefore, appetite-stimulating NPY/AgRP molecules inhibit the message that no more food is needed in two ways: They limit production of MSH and they prevent existing MSH from signaling.

This redundant system, then, promotes eating by (1) directly prompting food intake via the appetite-stimulating molecules NPY and AgRP and (2) sending signals that inhibit POMC neurons from suppressing appetite. Strikingly, even though the predominance of appetite-stimulating activity in this system is driven by negative energy balance— the need for food— its influence over POMC neurons seems to be the brain's default system. In other words, the human brain has the default wiring: "I am hungry."

This system had profound evolutionary consequences. The redundant mechanism to promote eating aided in survival when food was scarce, yet today when food is available at our discretion and the energy stores (fat tissue) are sufficient to maintain the energy needed for the body's metabolism, the brain's default wiring still signals us to keep eating. This mechanism, therefore, might be a key contributor to the current worldwide epidemic of obesity.

Enter Leptin and Visions of Treatment

More than a century ago, British neurophysiologist Sir Charles S. Sherrington suggested that certain factors arising from the blood regulated food intake.[3] Seven decades later, scientists discovered that a naturally obese mouse (named ob/ob because of its mutant genes) became leaner if it received some blood from a naturally lean mouse.[4-6] These experiments provided convincing data that something about the blood was driving the physical characteristics of the ob/ob mice. Since then, growing evidence has shown that the tissues in the brain and the rest of the body communicate, almost certainly by the way of the bloodstream, to signal and to sense stimuli that regulate food intake.[7]

In 1994, Jeffrey Friedman and others at Rockefeller University discovered that ob/ob mice lacked leptin, a hormone produced by fat tissue. Researchers later found that leptin enters the brain and signals the ARC neurons to decrease food intake, thus inducing satiety.[8, 9] Then, in 2004, innovative research revealed that leptin can induce a re-wiring of NPY/AgRP and POMC cells in the ARC in such a way that the network departs from its default signaling of hunger and adapts to a positive energy balance, thus decreasing appetite.[10]

Administering leptin to obese patients to decrease their appetites, however, proved ineffective. The leptin treatment cured only a few individuals with a rare, genetic type of obesity (due to mutations in the gene that produces leptin, as was the case with the ob/ob mice). Further investigation proved that, with the exception of patients with this rare genetic mutation, the brains of obese people somehow develop a resistance to leptin, and the cells in the brain stop sensing the levels of leptin in the blood.[11] Thus, because the brains of the great majority of obese people cannot sense levels of leptin, using it to treat obesity is ineffective from the start.

Understanding the causes of leptin resistance in obese people, though, may help researchers identify new targets for treatment designed to restore their sensitivity to this hormone. Finding ways to heighten tissue sensitivity to hormone signals is difficult but not impossible, as diabetes management shows: People with type 2 diabetes are treated with compounds to increase their sensitivity to insulin. We hope for the development of a similar drug treatment for obese people, but, until we know what causes their resistance to leptin, investigation into such new compounds will remain in its infancy.

Researchers have found, in addition to leptin, several other hormones and molecules that regulate food intake by acting directly in the brain.

For example, the gastrointestinal system has been a target of research because scientists believe it releases many hormones to signal a negative or positive energy balance. Investigators have taken a particular interest in the finding that the gut, mainly the stomach, produces the important appetite-stimulating molecule ghrelin, which also acts in the hypothalamus to prompt eating. Researchers found that ghrelin activates the NPY/AgRP appetite-stimulating neurons in response to negative energy balance (for example, when a person is fasting or following a low-calorie diet).[12–14]

Studies investigating how ghrelin affects the activity of NPY/AgRP neurons in the ARC have encouraged researchers by showing a cascade of events in these cells that ends by stimulating food intake.[15] This sequence of events suggests several points of possible intervention for new therapies to treat energy balance disorders. First, ghrelin released by the gut enters the brain and stimulates a receptor on the NPY/AgRP cells that prompts secretion of growth hormone, thus increasing their appetite-stimulating activity. At the same time and related to this event, the mitochondrial machinery in the NPY/AgRP cells speeds up to deliver enough of the high-energy molecule adenosine triphosphate (ATP) for cell metabolism. The enhanced mitochondrial activity also increases the production of harmful oxygen molecules derived from mitochondrial respiration. These harmful molecules are free radicals that react with other molecules inside the cell and promote cell damage (for example, damage to the DNA). To protect the cell from such oxidative damage, the NPY/AgRP cells call on another mechanism to buffer these free radicals by activating the mitochondria's uncoupling proteins. With appropriate buffering of free radicals by the uncoupling proteins, the NPY/AgRP cells can maintain high firing rates, thereby stimulating food intake.

Understanding this process may affect more than obesity research; it opens many lines of investigation to target cellular pathways to regulate energy balance—for example, by aiming at uncoupling proteins or the cells' methods for buffering free radicals. In addition to suggesting an approach to solving eating problems, the possibility that free radicals play a role in modulating appetite and regulating energy homeostasis provides a promising avenue for the development of treatments for many diseases related to metabolism. Researchers can now test this promising theory with several newly developed compounds that act as antioxidants.

Moreover, scientists have identified several other molecules involved in regulating food intake. We won't describe them, but they serve as a reminder of how intricately evolution built this regulatory network. For example, the gut produces several eating-related molecules in addition to ghrelin. Adipose tissue produces, in addition to leptin, immune system molecules with a role in diabetes and obesity. Lastly, classic regulatory hormones such as the glucocorticoids, produced by the adrenal glands and thyroid hormones, also control food intake.

The Feasibility of Treating Obesity with New Pharmacological Therapies

In the past two decades, researchers have worked hard to understand the biological abnormalities involved in obesity and to identify cellular pathways as possible targets for pharmaceutical treatments. The brevity of leptin's shining moment dashed the hope that a single molecule could cure obese patients by counteracting their positive energy balance. This may be due to the different types of fuel utilization and overall metabolic consequences of the activity of NPY/AgRP versus that of POMC neurons. Fatty acids drive the firing of NPY/AgRP cells, while glucose drives the firing of POMC cells, showing that the former, which promote feelings of hunger and stimulate eating in response to negative energy balance, are a priority for survival.

Add this dominance—not to mention the default status—of the "get something to eat" component of the melanocortin system to the evolutionarily programmed rapid adaptability of brain circuits in response to the changing metabolic environment, and you can understand why it is a daunting and futile, if not counterproductive, task to attempt to develop a pill that will keep people from feeling hungry. Although a one-pill solution to obesity is unlikely, several avenues of research raise hope for new treatments for this widespread medical condition.

The first important step in managing obesity is to integrate disciplines. Based on the knowledge of the neurobiological basis of food intake, scientists could design a treatment using a mixture of compounds given at appropriate times. However, because of the influence of higher brain functions on the regulation of appetite—for example, the influence of the smell, taste and appearance of food on the stimulation of hunger—practitioners must also consider a psychological approach to treating obesity.

Indeed, we believe that obesity, like other disorders of energy metabolism (see "Anorexia Nervosa: a Mortal Clash between Reward and Hunger"), should also be treated as a psychological/psychiatric disorder. Additionally, because obesity involves not only elevated energy intake but also decreased energy expenditure, an exercise program is mandatory to its treatment. Finally, because obesity develops over many years, obese patients should expect a similar time scale to return to their ideal body weight after starting treatment.

This combination of treatment would take advantage of our new understanding of the brain's role in obesity, going beyond the "silver bullet" visions that followed the discovery of leptin. As scientists continue to learn more about appetite stimulation and appetite suppression in the brain, as well as the hormones and other players that contribute to the process, we trust we will soon reach the goal of stemming the spread of obesity.

Anorexia Nervosa: A Mortal Clash Between Reward and Hunger

Few disorders reveal the power of the brain's cognitive circuitry more clearly than anorexia nervosa, a psychiatric disorder characterized by extreme undereating, loss of body weight, hyperactivity and hypothermia. Compared with other psychiatric conditions, this disorder has the highest mortality rate. We theorize that, in cases of anorexia nervosa, the brain's ancient evolutionary wiring for adapting happily to low food availability is inappropriately activated and finds itself in a life-threatening battle with other brain signals demanding action to obtain nourishment.

One clue to the intensity of this clash is the elevated level of physical activity in patients with anorexia nervosa, a symptom that people have reported for more than 100 years. Several studies have established a relationship between obsessive-compulsive characteristics and exercise frequency in women with strenuous daily exercise routines and in hospitalized female patients with anorexia nervosa.[16] In the patient group, preoccupation with weight was associated with both the frequency of exercise and pathological attitudes toward it. Addictive and obsessive-compulsive personalities contributed to excessive exercise because of their obligatory, pathological thoughts promoting it. Among anorexia nervosa patients, those who exercise excessively have more bulimic symptoms, higher levels of general psychopathology about eating and a greater degree of body dissatisfaction, anxiety, somatization (physical symptoms with a psychological origin), depression and irritability.

Scientists view the tendencies toward mental alertness and continued normal-to-high activity levels (despite insufficient nutrition and weight loss) as being relatively unique to anorexia nervosa patients, versus individuals who experience semi-starvation due to causes such as illness, chemotherapy or famine. For both of these tendencies, the most plausible explanation is activation of evolutionarily old circuitry leading to reward upon reduced energy intake.

A final clue is another characteristic of anorexia nervosa patients: 90 percent are women, mainly in their late teens. This leads us to propose that a cellular mechanism, in association with the changing hormonal milieu that is characteristic of anorexia nervosa patients, unifies and orchestrates activation of key brain circuits, which in turn leads to the behavioral and endocrine manifestation of anorexia nervosa. Our hypothesis is that anorexia nervosa occurs following shifts in the circulating hormones ghrelin, leptin and estradiol, which alter key groups of neurons. These alterations bring about sex-specific structural and functional changes in particular circuits of the midbrain that transmit the chemical dopamine to communicate. Dopamine then triggers a reward response in the prefrontal cortex and hypothalamus to undereating and overexercise.

We further hypothesize that rolling back this shift in reward response could reverse anorexia nervosa, and that either eliminating ghrelin signaling or suppressing the number of available long-chain free fatty acids in the brain could accomplish this. Neuronal cells normally activated by ghrelin use these acids for energy; thus, eliminating the fatty acids would silence the ghrelin-activated neuronal population. Patients who received controlled leptin and estrogen replacement therapy also might see their anorexic symptoms diminish. Moreover, we predict that if doctors help at-risk patients maintain estradiol and/or leptin levels during the initial phase of disease, the patients will be less likely to undergo the shift in reward responses that leads to anorexia nervosa.

12

Vitamin D and the Brain

More Good News

By R. Douglas Shytle, Ph.D., and Paula C. Bickford, Ph.D.

R. Douglas Shytle, Ph.D., is an associate professor of neuroscience at the Center of Excellence for Aging and Brain Repair and the Archie A. Silver Child Development Center at the University of South Florida Health Sciences Center. Dr. Shytle has published numerous papers in the area of translational neuroscience, most recently on the anti-inflammatory and therapeutic properties of plant polyphenols for Alzheimer's disease.

Paula C. Bickford, Ph.D., is a professor of neuroscience at the Center of Excellence for Aging and Brain Repair at the University of South Florida Health Sciences Center. Dr. Bickford is also a research career scientist at the James A. Haley Veterans' Hospital in Tampa. Dr. Bickford's studies have advanced understanding of the effects of fruits and vegetables that are high in antioxidants on age-related deficits in learning and memory and on other biochemical and physiological markers. She has dedicated her career to the study of aging and the role of inflammation and oxidative stress. She was the president of the American Aging Association in 2002 and has served on its governing board off and on for the past 15 years.

According to a growing body of research, vitamin D has many roles in regulating brain health, from aiding the development of the brain and nervous system to postponing decline toward the end of life. R. Douglas Shytle and Paula C. Bickford review the research and argue that while it is clear that many people worldwide experience vitamin D deficiency, we need to complete much more research to understand fully how this deficiency affects brain health.

INTEREST IN VITAMIN D has surged recently for two reasons: So many people do not have enough in their bodies that vitamin D deficiency is now recognized as a worldwide health problem, and at least 100 reports suggest that vitamin D can prevent or treat an array of chronic medical conditions.[1] However, in light of the recent controversies surrounding the efficacy—and even safety—of other vitamins (in particular, vitamin E), it is important to evaluate the current research carefully, avoiding the temptation to allow another popular frenzy to polarize us, either in support for or against the use of a particular supplement.

While the current controversy regarding the use of vitamin E is beyond the scope of this article, we must remember that much of the confusion stems from the conflation of the large body of evidence supporting the preventive health benefits of eating foods rich in diverse natural forms of vitamin E (known as tocopherols) and the recent short-term studies investigating the use of a single synthetic form (known as alpha-tocopherol) to treat people with chronic medical conditions. We hope that everyone will learn from the mistakes made during the vitamin E craze and avoid promoting premature conclusions regarding the health benefits of vitamin D or any synthetic form of it.

Our bodies do not synthesize vitamins. Rather, we must acquire these small natural substances that through the foods that we eat. Vitamin D, famously known as the sunshine vitamin, is not really a vitamin but a hormone, since our bodies produce most of it when we expose our skin to the sun. While it is well known that vitamin D deficiency causes rickets in children and can lead to osteoporosis and bone fractures in adults, recent research suggests that such a shortage is associated also with an increased risk of various types of common cancers, diabetes, autoimmune diseases, hypertension, stroke, infectious diseases and psychiatric illness. It is rapidly becoming clear that vitamin D plays various roles in regulating optimal brain health, both during development of the brain and the nervous system and throughout our lives.

Vitamin D and Early Brain Development

Until recently, scientists thought that only certain peripheral organs could synthesize the final active form of vitamin D. But we now know that human and rodent brains do express the protein necessary for the conversion of vitamin D to its final active form, and that the nutrient binds to sites on brain cells in a similar pattern in both humans and rodents. This evidence suggests that vitamin D's roles during normal brain development are so important that they have been preserved through evolutionary change. It also suggests that we might learn more about humans by observing the rodent brain.

Growing evidence from a group of studies on both rats and mice indicates that vitamin D is involved in normal structural brain development, though it is not yet clear whether that is the case in humans.[2] Mice born to mothers that were deficient in vitamin D before and during pregnancy had longer, thinner brains with enlarged ventricles (brain fluid canals). These offspring grew more new brain cells than normal during early brain development and had less "pruning" (death) of excess cells, a necessary process for forming effective brain cell connections. These effects may be permanent: Some rodent studies report that vitamin D supplementation after birth does not reverse these alterations.

In another series of studies, mice genetically engineered to lack cellular vitamin D binding sites, so that their brain cells do not receive the nutrient, appear to have an impaired response to novel environments, similar to the response observed in people with autism or schizophrenia. While this is only a first step toward characterizing the behavioral and brain abnormalities of offspring born to vitamin D–deficient mothers, it is an important finding. Epidemiological studies suggest that there is an increased risk of schizophrenia and autism in populations with relatively less sun exposure and vitamin D production, such as people living at high latitudes and babies born during winter months.

These findings are associations, not examples of cause and effect. Future studies should aim to clarify these correlations, especially since roughly half the world's pregnant women and newborns currently have insufficient blood levels of vitamin D—even in populations that experience normal sun exposure. In particular, more research should be directed at investigating the hypothesis that vitamin D deficiency contributes to the apparent increase in cases of autism, as measured by the increase in diagnosis of the disease among children. The increasing number of autism diagnoses in the past 20 years appears to correspond with the increasing use of sunscreens and sun avoidance, which may have led to lower vitamin D levels in developing brains.

While we still have much to learn about how vitamin D acts in early brain development, a likely hypothesis is that it regulates substances known as bone morphogenetic proteins. These proteins are important in triggering stem cells to differentiate into many types of cells, including brain cells. Evidence suggests that vitamin D increases the production of these proteins. In addition, the proteins' signaling pathways may serve an essential purpose during brain development, blocking brain cell division.[3] This cell-division process appears to be missing in vitamin D–deficient mice. However, no research group has yet conducted the pivotal experiments that would connect these dots with assurance.

Vitamin D and the Adult and Aging Brain

Recent epidemiological studies report that a deficiency of vitamin D appears to raise people's risk for fatal stroke, dementia and multiple sclerosis (MS). For example, in a recent study, Thomas Wang of Harvard Medical School in Boston followed 1,739 people (at an average age of 59) for five years.[4] Participants with low vitamin D levels had about a 60 percent higher risk of experiencing a cardiovascular event, such as heart attack or stroke, than participants with higher levels of the nutrient, even after accounting for other well-known cardiovascular risk factors such as diabetes, high cholesterol and high blood pressure. Participants with both high blood pressure and vitamin D deficiency had twice the risk for heart attack, heart failure or stroke.

Investigators at Heidelberg University in Germany reported similar results.[5] Of 3,316 people referred for evaluation of their heart arteries, those with low levels of vitamin D were more likely to have a fatal stroke in the next seven years (the median follow-up period), even after accounting for other cardiovascular risk factors. During studies on animals, these authors noted that vitamin D thins the blood and seems to protect neurons; the researchers' findings suggested that people who have had strokes or are at high risk for stroke should take vitamin D supplements. Findings from studies in which researchers induced strokes in animals also support the claim that a certain level of vitamin D can prevent or treat stroke.

The results of these studies are consistent with psychiatric research investigating the cognitive and mental impairment associated with the age-related decline in vitamin D levels referred to as hypovitaminosis D (HVD). At least 40 percent—one study indicated 90 percent—of older adults, including people who live in sunny areas such as Florida, have HVD.[6,7] Because vitamin D is fat soluble, many elderly people (who often have a higher fat-to-muscle ratio) may retain more vitamin D in their fatty tissue and thus have less of the nutrient available in the blood to maintain proper health. In addition, as we age, our skin becomes less efficient at making vitamin D in response to the sun.[8]

A recent review of the research on vitamin D deficiency and mental disorders by Paul Cherniack's group at the University of Miami found five studies reporting an association between HVD and dementia, four studies linking HVD to mood disorders such as depression and bipolar disorder and four studies linking HVD to schizophrenia.[9] Only two studies to date found no relationship between HVD and mental illness: one study of depression and one study of dementia.

The bulk of the evidence so far suggests that people who have HVD are at greater risk for conditions such as stroke, dementia and mood disorders than those who do not. What remains unclear is whether short-term oral vitamin D supplements would reverse these conditions in people who already have them. Again, more research is needed.

A critical issue in designing future clinical studies is defining the optimal dosage of oral vitamin D. For example, Reinhold Vieth, a prominent vitamin D researcher from the University of Toronto, has argued persuasively that clinical trials should use oral vitamin D doses equal to or higher than 800 international units (IU) per day (IU is a measurement based on a vitamin's biological effect), since most previous studies reporting clinical benefits of vitamin D for bone health used doses that exceeded this level. Already, in one trial that used only 400 IU per day (the current U.S. recommended daily allowance for adults, also described as 5 micrograms), participants with depression failed to show any benefit from the treatment.

One of the most active areas of vitamin D research is its potential connection to MS. As is the case with autism, the number of people with MS is higher in northern latitudes. Since vitamin D is produced primarily via exposure to sunlight, and reports have correlated high serum levels of vitamin D with a reduced risk of MS, researchers hypothesize that vitamin D may help protect people from the disease.

A recent study conducted by scientists at the University of Oxford and the University of British Columbia is being hailed as a major step in proving this hypothesis because it links the environmental risk factor of low vitamin D levels to a previously known genetic abnormality common to many people with MS. These researchers discovered that if a person has low levels of vitamin D, this gene does not function properly, thus making a person vulnerable to environmental triggers suspected in MS.[10] At least two current clinical trials are investigating the potential benefits of using vitamin D supplements as a treatment for MS.

We find compelling the scientific evidence from animal studies that vitamin D supports healthy brain function in general throughout life. Vitamin D appears to be a multipotent brain-cell-protective hormone; that is, it works through diverse and complex mechanisms, including brain calcium regulation, the reduction of oxidative stress, immune system regulation and enhanced brain cell signaling.[11]

More May Be Better

Between 20 and 80 percent of the world's population has insufficient vitamin D levels, and what this means for their brains and bodies is not yet clear. Why such a deficiency? Likely this has happened because we are hiding from the sun—the major source of vitamin D for most people. Very few foods naturally contain vitamin D, and foods fortified with the nutrient often do not have enough of it to satisfy our daily requirements. For example, a fair-skinned young adult who wears a bathing suit and spends 5 to 15 minutes in the midday summer sun can produce around 10,000 IU of vitamin D. To obtain the equivalent amount from his diet, he would need to drink 100 glasses of milk (at 100 IU per 8-ounce glass) or take 25 vitamin tablets containing the current recommended daily allowance (400 IU per tablet).

Considering that vitamin D supplements are low both in cost and in known risks, and that research suggests that vitamin D may benefit early brain development and ongoing functioning, we would like doctors to advise their healthy patients to optimize vitamin D production via the skin during summer months and to use vitamin D supplements and to eat more foods rich in vitamin D, such as fish, in the winter months (or year-round in the higher latitudes).

According to the available evidence, humans require 5 to 15 minutes of unprotected daily summer sun exposure (depending on skin type and proximity to the equator) to optimize vitamin D synthesis from the skin. We might consider 5 to 15 minutes of unprotected summer sun exposure, followed by exposure with sunscreen for the rest of the day to help mitigate the bad effects of prolonged exposure, such as the risk of skin cancer and premature skin aging. (Sun exposure also provides other health benefits for the brain that may not be directly related to vitamin D production, including improved regulation of mood. These are important topics for another paper.)

Daily oral vitamin D supplements can completely compensate for lack of vitamin D production from the sun. However, most studies suggest that the recommended daily allowance of this vitamin be increased to somewhere between 800 IU and 2,000 IU per day in order to provide maximum benefit to most age groups. Researchers recommend even higher doses for people with extremely low vitamin D blood levels.[12]

Two forms of vitamin D are available for supplementation. Throughout this article we have been discussing the animal form known as vitamin D3 (cholecalciferol). This form is the most desired form for supplementation and is available over the counter. Another form, derived from plants, is known as vitamin D2 (ergocalciferol). Although vitamin D2 is available by prescription, a growing body of evidence indicates that it is nearly ineffective as a vitamin D supplement and should be avoided.[13]

Several randomized, prospective, controlled trials among frail elderly people strongly suggest that these participants benefit from daily oral supplements of 800 IU or more of vitamin D, which enhances muscle strength and decreases risk of bone fractures. The current recommended daily allowance for elderly people, 600 IU, is not enough, the research suggests; increasing the dosage in these individuals, particularly those who are inactive (housebound or institutionalized), may prevent or ease some symptoms of frailty. Also, it appears that there is a wide safety margin in dosage—recent research has detected signs of vitamin D toxicity only after daily doses exceed 50,000 IU for several days or weeks.[12] Thus, doubling or even tripling the standard vitamin D supplement doses or fortifying more foods with vitamin D would be very safe.

Such proposals may sound alarmingly similar to the hype that flooded the media when the vitamin E craze started. As with other new findings, people should discuss this new research on vitamin D with their health care providers before beginning to take large doses (greater than 2,000 IU per day) of vitamin D on their own, especially if they have a medical condition or are taking other medications that may interact with the nutrient. Another good source of information is pharmacists, who are often especially aware of new preventive health guidelines, drug-supplement interactions and what dietary supplement products have the best quality standards.

To confirm the nutrient's benefits in humans, we would like to see well-controlled clinical trials of vitamin D supplementation in people who have clinically low vitamin D blood levels. Such trials seem likely to have low risk and high return in terms of preventing disease and promoting health.

Long-term public health goals should include validating optimal levels of vitamin D, especially in women before pregnancy, and working to maintain appropriate vitamin D levels throughout everyone's life span. After the damage is done, we cannot rely on the potential short-term benefits of using vitamin D supplementation or, worse, synthetic vitamin D analogues (as we did with vitamin E).

13

Religion and the Brain

A Debate

By Dimitrios Kapogiannis, M.D.,
and Jordan Grafman, Ph.D.;
by Andrew B. Newberg, M.D.

Dimitrios Kapogiannis, M.D., is a staff neurologist at the National Institute on Aging. He studied medicine at the University of Athens and specialized in neurology at Massachusetts General Hospital in Boston. He specialized in cognitive and behavioral neurology as a clinical fellow at the Cognitive Neuroscience Section of the National Institute of Neurological Disorders and Stroke. His research interests include the neural basis of religion and personality.

Jordan Grafman, Ph.D., is a senior investigator at the National Institute of Neurological Disorders and Stroke (NINDS), part of the National Institutes of Health. He has served as chief of the Cognitive Neuroscience Section of NINDS since 1989. He is an elected fellow of the American Psychological Association and has received both the Defense Meritorious Service Award for his studies of brain-injured Vietnam veterans and the National Institutes of Health Award of Merit.

 **Andrew B. Newberg, M.D.**, is an associate professor in the department of radiology and psychiatry at the Hospital of the University of Pennsylvania, where he also is the director of the Center for Spirituality and the Mind. Newberg is the author of *How God Changes Your Brain* and *Why God Won't Go Away: Brain Science and the Biology of Belief.*

Does evolution explain why the human brain supports religious belief? Dimitrios Kapogiannis and Jordan Grafman, scientists at the National Institutes of Health, follow up on a recent scientific paper by stating that brain networks that evolved for other purposes have given rise to our capacity for religious belief and experience. Andrew Newberg, the radiologist and psychiatrist who wrote How God Changes Your Brain, *takes a different approach. He argues that the brain may be an instrument of religious experience but is not necessarily the origin of that experience. Each side of the debate first wrote a position statement; the sides then exchanged statements and wrote rejoinders.*

How Our Brains Evolved to Accommodate Religious Belief

Dimitrios Kapogiannis and Jordan Grafman's opening statement

Every school of philosophical thought has proposed its own account of how religious belief originated. Philosophers typically consider religion to be a cultural and historical phenomenon without a foundation in science. They neither attempt to bridge different approaches to religion—psychological, cognitive, behavioral, social, political and historical—nor distinguish among religion's different aspects, such as belief, experience and ritual, in a way that enables people to test concrete hypotheses. However, recent progress in understanding the neurobiology of social cognition has opened the door to a neuroscientific perspective on religion.

Scientific explanations for complex biological phenomena are not reductionist. Rather, they require synthesis of the various components and their interactions at different levels. To explain religion in biological terms, therefore, we need to define both its characteristics in an individual and the variability of its expression among people and cultures.

Religions and their accompanying belief systems are cultural universals. Relying upon cultural evolution alone to explain this ubiquity requires acceptance that the innovation of religion transpired at the dawn of human history and *all* human societies have perpetuated it separately, which seems highly unlikely. Moreover, we now know that other evolutionary phenomena, such as symbolic language and morality, have solid bases in biology and information processing.

Many current theorists regard religion as either an evolutionary adaptation or a byproduct of certain adaptive changes, driven in either case by the development of larger social groups and more complex interactions among them. These theories link the emergence of religion in our ancestors with the development of cognitive processes: theory of mind, the ability to interpret the intentions and emotions of others; social cognition, or neural processes concerned with such social phenomena as morals and group identity; intuitive (prescientific) theories about natural phenomena; causal reasoning; and symbolic language. These cognitive processes have different evolutionary origins, and presumably they resulted from the expansion of specific brain regions. Indeed, our research involving functional brain imaging of the invoking of religious beliefs leads us to conclude that religion emerged as a combination of cognitive functions, the main evolutionary advantage of which was probably unrelated to religion.

In an individual, the term *religiosity* refers to a cluster of personality traits related to the adoption of religious beliefs and engagement in behaviors reflecting those beliefs. Due to both environmental and genetic factors, degrees of religiosity vary widely among modern humans. From an evolutionary standpoint, the variety stems from a lack of selection pressure—no single set of beliefs and associated behaviors conferred a survival advantage relative to others. As an evolutionary adaptation, religiosity resembles language, which humans adapted for social communication. The evolution of linguistic ability in the ancestors of modern homo sapiens clearly occurred at the biological level, and this evolution is a hallmark of modern humans. Fossil records reveal a gradual increase in the size of brain areas critical for language over tens of thousands of years. When groups of biologically nearly identical modern humans became geographically and socially separated, individual languages—like discrete religions—emerged and acquired their own *cultural* evolutionary histories (with a rate of change higher than Darwin's theory of evolution would predict for biological traits). These distinct histories result from an accumulation of seemingly random changes, but also from the selection of features that conferred some advantage, such as languages' differential prevalence of vowels and consonants based on climate.

Yet virtually all human beings have a comparable capacity for language, while the capacity for religion appears to be highly variable. Among our predecessor primate species—or groups within them—natural selection must have extinguished those with language deficiencies. In contrast, there are people with no supernatural beliefs—at least in the Western world, where alternative theories about how the world was created and how it evolved are widespread. It appears, therefore, that because natural selection did not eradicate populations that did not hold religious beliefs—or did not strongly adhere to them—there can be a high degree of variability in modern populations with regard to religion.

Brain Networks Involved in Religion

What, then, is the neurobiological basis of the highly variable human belief system? We found evidence that well-characterized brain networks are involved. Despite seemingly daunting differences, we organize religious belief around three principles, or dimensions, at the cognitive level—at least among members of Western societies—and both religious and non-religious people share these organizing principles. A secondary process, then, determines an individual's specific expression of his or her beliefs. Researchers previously have implicated these neural circuits in understanding others' actions, intents and emotions, as well as in processing abstract language and imagery.[1] These basic cognitive and social skills are prerequisites for developing a sophisticated religious belief system.

In particular, the evolution of brain networks concerned with understanding the actions of others seems to have made possible concepts of a godlike entity's involvement in human life. The crucial brain areas for this function are in the part of the frontal portion of the brain that also is involved in observing purposeful human action and detecting underlying intentions. These brain areas work with other regions to decode the emotional impact of the actions we observe.

A self-centered analysis of complex social interactions must have been crucial not only for the survival and status of an individual among larger social groups, but also for the evolutionary stability of these groups. An individual's emotional life includes decoding others' emotions and employing them in association with his own goals. Moreover, regulating emotions—through such skills such as deception, for example—optimizes social performance. Our research demonstrates that a person's sense of love and anger from a godlike entity derives from these social functions.[1] This sense is based in brain areas whose evolution enabled us to detect emotion from others' facial expressions and tones of voice, as well as attribute personal relevance to social phenomena.

The previous two dimensions—understanding others' actions and intents and decoding their emotional impact—encompass perceptions of the level of involvement and emotion of God or another supernatural entity in the construction of religious belief. The third dimension refers to the source of religious knowledge—what individuals have learned and experienced. This final dimension, we propose, influences how our brains code beliefs and connect them with other sources of knowledge. Together, the three dimensions we have identified help individuals construct religious belief systems that interact with other belief systems, social values and morals to help determine goals, control behaviors and balance emotions.

We should note that detecting another person's intent is perhaps the earliest (pre-linguistic) form of causal reasoning;[2] it allows us to predict future outcomes based on others' current behaviors. Perhaps, in early, prescientific attempts to explain physical phenomena or historical coincidences, our ancestors needed to imagine supernatural intervention. Children arrive at such default explanations at specific times during their development and sometimes hold on to them as superstitions throughout adulthood.

Such supernatural explanations may be reinforced by evolutionarily ancient neural networks that code rewards and punishments, and the uncertainty regarding expected rewards and events we find threatening.[3, 4] In a danger-laden world, such as the one in which our ancestors evolved, the human brain may indeed have coded as a reward any explanation minimizing fear or the uncertainty of threats,[3, 4] and this coding might even have offered a survival advantage.[5, 6] A coherent world theory that assumed the existence of a supernatural being or beings may thus have had survival value at the individual level. Furthermore, adoption of such explanations by members of a group may have increased the predictability of their behavior, defined and signaled group membership and, therefore, promoted cooperation and had survival value at the group level.

The complexity of social interactions in these larger groups required abstract symbolic coding of ideas and mental states, and thus paved the way for symbolic language to evolve. This complexity also required people to mentally simulate possible social scenarios and outcomes, which supported the evolution of mental imagery (an ability that, in turn, promotes learning, even at the elementary level of motor imagery). These abilities, along with the associated brain areas, enabled humans to develop a wide variety of religious and other beliefs. Doctrine, which refers to beliefs that are transmitted culturally rather than grounded in personal experience, is a special type of abstract idea; it engages brain areas involved in the processing of abstract language.

Another piece of the puzzle is the key involvement of visceral emotions that occur in both social interactions and religious behavior. In the course of human evolution, basic emotions such as disgust and fear acquired new social equivalents such as moral outrage and guilt. Religious practice successfully engages these social emotions. We have shown that, when devout people disagree with certain religious beliefs, activity increases in the brain's anterior insular cortices—areas involved in disgust, aversion, guilt and fear of loss.

More Than a Primitive Response

We conclude that there is nothing special about the source of religious knowledge or the brain networks involved. In the brain, religious knowledge relates to, and may be vulnerable to modification by, other sources of knowledge. These neural connections could account for the historical observation that religious ideas tend to cluster with certain political or social ideas more than we would expect simply from a random co-occurrence—an observation suggesting that religious ideas could be subordinate to a higher-order classification of concepts.

Critics might seize upon our findings as evidence that religion is a phenomenon of the primitive mind, and it might one day disappear as science "enlightens" humanity. Not so fast: Our need for religion might be embedded in our biology. Religious belief engages some of the most recently evolved brain areas, which perform uniquely human functions that define our species: the ability to comprehend the intentions and feelings of our fellow humans, symbolic language, reasoning. For better or worse, humans are not strictly logical creatures but social animals. We imagine, observe, interpret, love, and occasionally detest each other. Therefore, we cannot consider religion strictly an outdated response to the modern world.

Instead, we believe that religious belief emerged for the purpose of social structure. Social structure originally was based upon principles derived from small family, group and tribal social interactions and a need to explain natural phenomena that did not appear to have an obvious human or animal physical cause. Then, as societies grew larger, religious belief further developed through the establishment of greater religious infrastructure. This emergence and adaptation of religious belief depended on the sophisticated cognitive and neurobiological processes we have described. In addition, if human brain evolution gave us *foresight* as a weapon against stronger foes and natural phenomena, then religious beliefs that concerned an afterlife might have been an effort to extend the boundaries of life in a way that was consistent with this newly found ability.

Although we have rightly ceded explanations for natural phenomena to science, we still struggle to create optimal social relations within and among societies, and in this quest, religion continues to play a vital role.

Religion, Evolution and the Brain: What Caused What?

Andrew Newberg's opening statement

Where did religious and spiritual beliefs come from? The answer to this question depends on your own belief system. The position of some people who are not religious echoes Sigmund Freud and, more recently, Richard Dawkins: Religion is primarily a pathological mistake made by the brain. Others with a less negative view consider religion to be a constructive creation of the brain. People holding the latter view might claim that evolutionary forces affected the human brain in such a way that it created religion as a means to better adapt to the world around us. Can evolution explain why the human brain supports religious beliefs? I argue that although explanations that focus on how brain structures and functions have evolved may provide important information regarding the raison d'être of religion, this "neuroevolutionary" approach can be limited.

One problem with this approach to religion is the difficulty in discerning the element or elements that are adaptive—that undergo change to enhance the probability of survival. For instance, different models have focused on the sense of control over the world that religion helps us to achieve, religion's provision of social cohesiveness and moral foundations, its potential physical and mental health benefits or its utility in providing answers to questions that we cannot fathom. Still other theorists cite the importance of religious and spiritual *experiences* as primary evolutionary sources of religion.

A religious perspective challenges all of these neuroevolutionary approaches by reversing the causal arrow's direction: Perhaps religious belief causes the brain to change rather than the other way around.

A religious individual looks outward for religion's origin. Thus, the most common answer is straightforward: Religion comes from God. For a religious individual, it is no surprise that religion and spirituality are a part of the human brain—a God who provided human beings with no physiological way of having any kind of relationship with God would leave us with a fundamental theological problem. This explanation holds that religious beliefs originate with God, but thereafter, the human brain takes over to determine how we manifest those beliefs in our religious and spiritual practices. So, while an understanding of the brain may help us better comprehend how we become religious or spiritual, the brain only constrains or directs us toward those beliefs; it does not create them. This argument also helps explain why each religion has a different perspective on the meaning and nature of God, particularly God's relationship to human beings.

We can question the validity of the religious explanation—which clearly argues against a neuroevolutionary cause of religion—because there are no scientifically derived empirical data to support it. How, then, do we know which explanation is correct? The fundamental problem is in evaluating how the brain perceives and understands reality. This dilemma forces us to re-evaluate what constitutes absolute fact and consider the potential need for an integrated epistemological approach to the question of how we know what is real.

The difficulty we face is how to evaluate the validity of different perspectives on the origins of religious and spiritual beliefs. Members of the emerging discipline of neurotheology—the study of how spiritual experiences and neural processes affect one another—are attempting to address this quandary by striving to combine neuroscience data with religious and theological ideas in order to better understand the intersection of religion and neuroscience. Neurotheology differs from other approaches to neuroscience in that it maintains a strong foothold in religious and spiritual beliefs. Thus, neurotheologians do not necessarily attempt to *explain* religion exclusively on the basis of neuroscience. Religious thinkers might have some things to say about neuroscience as well.

Ultimately, neurotheologians should both maintain and take into account religious and spiritual doctrines, practices and experiences while upholding appropriate scientific rigor. Trying to strike this balance raises fascinating and challenging methodological issues. So, while some of my arguments might sound more rational than others, depending on your belief system, it is important at least to reflect on each of the perspectives before reaching any conclusions about such a complex subject.

Scientific Approaches to Religion

When we evaluate evolution-based theories and other perspectives on religion, we must address several methodological concerns. Many scientific approaches explore religion; each can lead to a different conclusion about religion's nature and origin. Therefore, even after we avoid the major temptation to explain away religion because of the lack of scientific evidence, methodological complications hinder our quest to make rigorously derived conclusions supporting an evolutionary basis for religion.

The Neurophysiology of Spiritual Practices

One scientific model for studying the origin of religion employs brain-imaging technologies to explore the physiological changes associated with a spiritual practice such as prayer or meditation. For example, using positron emission tomography (PET), single photon emission computed tomography (SPECT) and functional magnetic resonance imaging (fMRI),[7] researchers derive simultaneous measures of biological changes in the brain, including cerebral blood flow and metabolism, and electrical and electrochemical (neurotransmitter) activity. Investigators use subjective measures to assess each participant's psychological and spiritual feelings or thoughts, and then they compare the biological and subjective measures. Researchers evaluate additional physiological measures such as blood pressure, body temperature, heart rate and galvanic skin response (a measure of autonomic nervous systems activity) because these are frequently associated with brain changes, and previous research has shown that religious and spiritual phenomena affect body physiology.

The ideal result of these procedures would be a detailed portrait of brain activity correlated with a particular religious or spiritual experience. Such research has indeed helped to delineate the physiological correlates of such experiences, but physiological correlates by themselves cannot explain origin and nature—in other words, we cannot conclude that the brain activity is the specific *cause* of religious experience. Most studies have shown that multiple brain areas are involved, which complicates the ability to identify one or two physiological mechanisms that explain religion.

Other problems are more fundamental. Most important, it is difficult to assess whether the brain generates or simply receives certain types of experiences, such as the feeling of being in God's presence. A brain scan shows associated changes but does not demonstrate whether these changes caused the experience or were produced in response to an external stimulus.

Furthermore, researchers typically cannot obtain the psychological and spiritual data during such an experience, since that would require interrupting it. Even one tap on the shoulder to ask a research participant how he felt at that moment would destroy the occurrence we are trying to study. Thus, we can never be certain exactly when an intense religious experience actually occurred during an imaging session.

Finally, subjective measures typically are based on participants' responses to questions about what they felt, thought or perceived during the experience, but these responses, reflecting cognitive processes, are not necessarily the basis of a true spiritual episode. An inherent scientific bias in such studies is that investigators are measuring nothing more than cognitive processes of thought, feeling and experience, rather than something inherently spiritual (whatever that means from a scientific perspective).

Creating or Altering Spiritual Experiences

A second scientific method for studying the origin of religion involves trying to alter a participant's religious and spiritual experiences. This approach might employ the use of drugs to directly affect or stimulate a spiritual experience. Because certain hallucinogenic drugs and stimulants can induce spiritual experiences, careful research, perhaps utilizing modern imaging techniques, may help elucidate which neurobiological mechanisms are involved. Researchers already have investigated the use of such hallucinogenic agents, but more extensive study, particularly related to religious and spiritual episodes, is necessary to gain a better under-standing of the range of their effects.[8] From a scientific perspective, one of the limitations of such studies is that different hallucinogens affect different neurotransmitter systems, thus making it difficult to determine whether any one neurotransmitter system is responsible for the drug-induced religious experience. Moreover, if multiple neurotransmitters are involved, how can we conclude which neural pathway—and hence, which evolutionary element—resulted in religion?

In addition, the role of drugs in many shamanic and native cultures turns the neuroevolutionary theory of religion on its head. For thou-sands of years these groups have used psychotropic compounds to induce spiritual states. But rather than conceiving of such effects as biological or artificial, these cultures see the drugs as opening the mind up to the spiri-tual realm. For them, drug use is not unlike putting on a pair of glasses to see more clearly. The drugs merely take the brain to another level where it can perceive the world in a clearer or, perhaps, higher way. From this viewpoint, the brain enables spiritual and religious phenomena rather than causing them. To put it another way, such cultures would think brain evolution an effect of the spiritual realm rather than a cause of it.

Spiritual Experiences Related to Brain Injury or Disorders

A third neuroscientific method for exploring spiritual and religious phenomena is to study patients diagnosed with neurological or psychi-atric conditions. For instance, studies have linked temporal lobe epileptic seizures, brain tumors, stroke and other brain injuries to spiritual expe-riences or alterations in religious beliefs. Temporal lobe epilepsy in particular has been associated with hyperreligiosity and religious conver-sions.[9] Psychiatric disorders such as schizophrenia and mania also have been associated with spiritual experiences and conversions. Delineating the type and location of the brain alterations involved in these conditions will help scientists explore the biological substrates associated with patients' spiritual episodes. However, clinical researchers must take care to avoid referring to spiritual experience only in pathological terms or as associated with conditions of brain disease or injury. This approach sometimes leads people to classify religion as delusional or abnormal because they define it only as part of a disease state.

In contrast, most religious individuals do not exhibit signs of a neurological or psychological disorder, and researchers have demonstrated that religion can help people cope with stress and, in many cases, reduce anxiety and depression. Thus, while psychopathological approaches provide a unique perspective on religious phenomena, they suggest that religion is not at all adaptive. This conclusion contradicts theories proposing that religion is an evolutionary process.

A Specific Focus on Brain Evolution

A more specific evolutionary approach to the study of religion typically focuses on two important aspects of human evolutionary development: social interactions and cognitive processes. Both appear central to religion.[1] Socially, it seems that human beings need to seek out and develop personal relationships that eventually can lead to the formation of a society on a grander scale. Cognitively, the human brain appears to continuously categorize and analyze the world to develop meaning and, perhaps more important, to determine appropriate social behaviors that will have survival benefits. Therefore, even though social behavior and religion appear to involve the same brain structures, we cannot assume that social behavior, rather than cognitive understanding and control, was the primary adaptive advantage that led to expansion of religion throughout the human world. Rather, any evolutionary advantage of religion could be multifactorial; it could include beneficial adaptations to social, cognitive, health, ethical and environmental factors.

The abilities to perceive and evaluate environmental dangers, to respond appropriately to situations and to weigh alternatives are critical to evolutionary adaptation. Our brains differentiate self from other, order things in space and time, perceive interrelationships among objects in the world and use symbols and language to express ideas. Religion may help to engage these cognitive processes, which we use to try to understand and control our world.

At issue is how each of us understands reality. Our individual perceptions of reality ultimately lead each of us to conclude whether religion is nothing more than a product of the brain (adaptive or not) or a necessary result of a spiritual realm that our brain may occasionally access. This fundamental epistemological problem challenges all aspects of human thought—scientific, philosophical and theological.

How Do We Know What's Real?

This epistemological problem may prove to be the ultimate challenge to a purely scientific understanding of religion. How do we know whether anything we perceive is real or not? Put another way, how do we know if the reality of the external world corresponds, at least partially, to our mental representation of it? Humans have posed the question of realness since the dawn of philosophy, science and religion, and the question has generated various answers.

This reality question lies at the very heart of neuroscience. If our understanding of the world comes from the brain, it is subject to a variety of misinterpretations, misperceptions and misunderstandings. How, then, are we to know whether the reality we perceive—scientific, religious or otherwise—is the true representation of the world? Ultimately, after much philosophical, theological and scientific exploration, we may be forced to arrive at the rather uncomfortable circular statement that what we take to be real is dependent only on how real it feels to us. From the neuroscientific perspective, this is consistent with a wide array of research showing how our brains construct our senses of reality.

We return at last to neurotheology as an approach to understanding the nature of all types of experiences of reality. Both science and religion provide potentially important information about the world that our brains perceive. We may ultimately find that religion is nothing more than a manifestation of the brain's function set in place by millions of years of evolution. We might find that perceived spiritual dimensions help us to get in touch with the more fundamental nature of reality. Either way, we should tread carefully and strive to understand reality—on all levels.

Response to Andrew Newberg

Dimitrios Kapogiannis and Jordan Grafman

We read Newberg's essay with an eye toward common ground. We whole-heartedly agree with Newberg that religion can help people cope with stress by lowering anxiety and also may provide an ethical basis for inter-acting with the world.

Our quarrel with Newberg's perspective is that he shies away from the scientific method's commonly accepted grounding in natural causes and effects, reproducible experience and logical reasoning. Our essay is a purely scientific dialogue: We did not seek to criticize the usefulness of experimental design, but instead explored whether religious beliefs are special compared with other belief systems by discovering the brain systems and cognitive/social processes involved.

Newberg raises a number of points that we think deviate from a rigorous examination of religion. For example, he argues that religious belief may "cause the brain to change rather than the other way around." But research demonstrates that almost all behaviors cause the brain to change via prac-tice and adaptation. Religious belief is not unique in that regard.

Furthermore, Newberg implies that exercise of brain functions over time wouldn't significantly influence the development of new knowl-edge or ways of being. Yet it is absurd to suggest that brain functions don't influence the development of belief systems. At the level of the individual, exercise of a brain function (say logical reasoning or imagina-tion) can have a profound influence over the range of beliefs she accepts. Similarly, at the level of society, promotion of the use of brain func-tions (say abstract language or empathy) can also change the prevailing pattern of beliefs.

It is also reasonable to speculate that, in an evolutionary time scale, gradual evolution of a range of brain functions enabled the emergence and adoption of myriad religious beliefs. Even modern biblical scholars and many religious practitioners would admit that there is little objec-tive evidence that God has completely scripted the requirements of religious belief. The hypothesis to challenge here should be that religious belief emerged out of the cognitive and social capabilities of humans and that those abilities depended upon the structure and function of the human brain.

Even if we could persuade Newberg that the above argument is valid, he still might argue that we need a special branch of cognitive neuroscience, including dedicated neurotheologians, to study religious belief, and he may be on solid ground here. Nonetheless, psychology has always had a theoretical and an applied component. We and others on the theoretical side work to determine the underlying principles of human behavior and neural functions, while those in the applied school assess how those basic principles relate to specific circumstances. The key to understanding both theoretical and applied findings is to maintain the link between the two and to identify analogies to results in other disciplines.

At times we found Newberg's statement confusing. For example, he wrote that it is difficult to assess whether the brain generates or *receives* (our italics) certain types of experiences, such as the feeling of God's presence. Let us be clear: It is simply a supernatural declaration to say that God has issued a stimulus to alert us to his presence without our having the ability to detect it with modern instrumentation. For better or worse, no scientific instrument ever designed by humans can detect God, and our findings suggest that we don't have a dedicated sensory organ or neural area dedicated to Him. On the other hand, many cognitive functions, such as imagination, do not have any obvious external causes and instead are generated internally. Scientifically, we can approach those functions only by correlating subjective experiences with objective measurements of neuronal activity. Here again, there is nothing special about the study of religion in the brain. It seems that Newberg only plays devil's advocate in raising the issue of the legitimacy of the neuroscientific methods for studying religion, since he bases his research on the same scientific principles as we do.

Newberg does not seem to observe the distinction between the evolutionary origins of religion as an almost universal human trait, which fits the timescale of biological evolution, and the origins of specific religions, which are better explained by cultural evolution. Moreover, he seems to wonder how it is possible for religion to have evolved for adaptive reasons, since much of the evidence for its neural correlates has emerged from the study of pathological states, such as epilepsy or schizophrenia. Again, there is nothing unique to religion, since our knowledge of physiology in general largely stems from studies of disease states. The deregulation of a mental or physiological function often provides the clues that lead to the understanding of its normal function. Lastly, Newberg refers to cultures that use drugs and shamans to explore religious belief and to reach ecstatic states. These are social phenomena worthy of documentation and study for their cultural effects and their impact on a person's experience and life view. Study of these agents or rituals likely will lead to the detection of brain regions important for these activities but not unique to them.

Science and religion may never reach common ground. Newberg seems to advocate a balance between incompatible reasoning systems. We have quite a different view of neurotheology, which we consider a branch of neuroscience that seeks to categorize and explain cultural phenomena based on tried and true neuroscience methods. Shamanistic cultures may offer different explanations for the origin of the world than does modern science; what we do not see is why this is relevant to the scientific study of religion.

In our view, religion constitutes a legitimate domain for scientific study, since the relevant phenomena are real and of great importance. We should determine the merit of any approach in terms of generalizeable knowledge. To quote Einstein, "Science without religion is lame, religion without science is blind."

Response to Dimitrios Kapogiannis and Jordan Grafman

Andrew Newberg

Kapogiannis and Grafman's research has provided another excellent piece to the puzzle of the nature of religiosity and religious belief. Their work

also provides an impetus for further study to uncover the biological correlates of religion. This has great importance for advancing and strengthening
research in the field linking neuroscience and religion, which I refer to as
neurotheology. Their scientific investigation helps clarify the brain regions
associated with specific components of religious belief.

In much of my own work, I have suggested that a large neural network
appears to be involved in religious phenomena, including experiences and a
vast array of beliefs. This model includes many of the regions Kapogiannis
and Grafman have identified, but their new research provides even more
detail. Given the richness and diversity of religious phenomena, which
Kapogiannis and Grafman appropriately point out, the brain network that
"gets into the act" is probably relatively large.

However, although experimentally defining cognitive and emotional
aspects of religion in the context of research is necessary for adequate
study, it also raises important concerns, as I have noted. In particular,
by pre-defining how religion makes us feel and think, we may end up
simply showing how the brain helps us feel and think in general rather
than discovering something that is truly unique to religion. In other
words, we might miss the part of ourselves that is inherently religious
or spiritual if all that we attempt to study is the cognitive neuroscience
of religion.

In terms of neuroscience, much of the research to date, including that
of Kapogiannis and Grafman, measures general physiological correlates of
religious phenomena. It also will be crucial to identify specific neurotransmitter systems that are involved in religious experience. This will likely be
the next step in evaluating the neurophysiology of religious phenomena.
And because many brain regions are implicated, researchers should focus
their attention on more than one neurotransmitter system.

Kapogiannis and Grafman's findings are consistent with previous
models of religious phenomena, which implicate parts of the frontal,
temporal and parietal lobes. As Kapogiannis and Grafman note, these
areas are involved in higher cognitive processes, social behaviors and
emotions. Such processes also play a critical role in religious phenomena.
It is reasonable for any neuroevolutionary analysis of religion to lead to

the conclusion that religion is built upon existing brain structures and their functions rather than on the development of a separate circuitry whose sole function would be supporting religious experience. Consistent with the findings of Kapogiannis and Grafman, there is no "God spot" in the brain. Rather, religion makes use of existing brain structures and their functions, and it appears that religious beliefs match up exceedingly well with those functions.

However, it is difficult to determine which of the functions related to religion ultimately provided the adaptive advantage that led religion to thrive throughout human history. Simply finding a relationship does not necessarily imply causality, and whether these findings ultimately imply that religion is nothing more than a brain-based phenomenon is another matter. The findings we are discussing link religion and the brain, but the brain may be receptive to religious experiences rather than creating them. Whether the brain generates religious belief or serves as a conduit for it remains a complicated question.

Book Reviews

14

Decisions Are Not So Simple

How We Decide
By Jonah Lehrer
(Houghton Mifflin Harcourt Publishing Co., 2009; 256 pages, $25.00)

Reviewed by Scott A. Huettel, Ph.D.

Scott A. Huettel, Ph.D., is associate professor of psychology and neuroscience at Duke University, where he is also director of the Duke Center for Neuroeconomic Studies. His laboratory uses behavioral and neuroscience methods to investigate the processes of decision-making, including both economic and social decisions. Central topics include the neural processing of risk and reward, strategic influences on decision-making, and individual differences in decision preferences. He is also lead author on the textbook *Functional Magnetic Resonance Imaging.*

THE BRAIN IS A SIMPLIFIER. When faced with a complex problem, it does not dispassionately weigh all available information before selecting a course of action. Instead, it takes in a situation, simplifies it and applies behavioral rules that have worked in the past. The brain's ability to make a quick decision based on limited information is often remarkably efficient—for example, when a driver swerves to avoid an unexpected obstacle. Yet we also encounter many dilemmas that the brain is less equipped to confront, from mundane situations ("Should I have this slice of pie for dessert?") to life-changing questions ("Should I quit my job and start my own business?"). For dramatic and sometimes disastrous examples of the consequences of poor decision-making, we need look no further than the current economy.

Jonah Lehrer's new book, *How We Decide*, explores the way people make decisions. It builds upon a recent surge in research in behavioral economics, cognitive psychology and neuroscience, much of which challenges long-standing economic models of rational decision-making. In the book's introduction, Lehrer sets forth his thesis: Our brains have a variety of tools for decision-making, some rational and others emotional, and good decisions come from applying the right tool in the right circumstances.

To support this thesis, Lehrer skillfully engages readers in a range of scenarios that emphasize the human character of decision-making. He introduces pilots faced with equipment failures, firefighters confronting an onrushing wall of flames, stock traders chasing market fluctuations, and Ikea shoppers bewildered by a vast selection of nearly identical couches. He then calls upon recent scientific research to explain the choices these people made. Across this diversity of content, Lehrer reveals his signal strength as a writer: an ability to link complex concepts without resorting to jargon or empty prose. He extracts meaning from a welter of complex phenomena and somehow produces a simple and coherent story.

Simplicity is not always a virtue, however, and complexity should not always be avoided. The task of the science popularizer, especially when relating cutting-edge topics like those Lehrer describes, is to simplify as much as possible without sacrificing fidelity to the original research.

Synthesizing the Brain: Reason vs. Emotion

Lehrer frames most of his examples within the dual-systems model of brain function, which echoes the ancient distinction between reason and emotion. As typically conceived, the dual-systems model suggests that our thoughts and behaviors reflect competition between two aspects of our brains. The rational system, often localized to the prefrontal cortex, supports controlled and sequential thought. The emotional system, which comprises regions such as the amygdala and the insular cortex, guides rapid, automatic and visceral reactions to stimuli. The moment-to-moment interaction of these two systems determines the extent to which our decisions are rational, emotional or both.

At first consideration, the dual-systems model seems like a reasonable account of human decision-making. In the wake of an impulse purchase or an extra piece of pie, we all have wished that we had exerted a bit more control. Given these shared experiences, many readers will find themselves nodding in self-recognition when Lehrer describes examples of irrationality: overemphasizing potential monetary losses compared with potential gains, seeing nonexistent patterns in stock fluctuations and seeking immediate pleasure at the cost of future pain. The dual-systems model allows us to ascribe these flaws to the limitations of our brains. In Lehrer's argument, the prefrontal cortex is rational but also easy to hoodwink, and our dopamine neurons, which respond to the rewards we receive, can do just that by detecting subtle and potentially misleading patterns in the world around us. Lehrer's conclusion: If only we could listen to our prefrontal cortices instead of our dopamine neurons!

Unfortunately for the lay reader, Lehrer's explanations of neuroscience lead to several problems. The first is his repeated tendency to shoehorn scientific findings into the overarching story. For example, Lehrer states that the dorsolateral prefrontal cortex, or DLPFC—the upper regions of the frontal lobes—is commonly considered to be "the rational center of the brain." He then tells a story about Mary, an Ivy League student who seemed to have a limitless future. She suddenly became profoundly self-destructive—absent from class, impulsive, alcoholic, promiscuous and unrepentant. An MRI scan revealed that Mary had a tumor in the prefrontal cortex. The natural conclusion is that the tumor had damaged the rational part of her brain. But hidden in this example is a subtle distortion: Self-destructive behavior such as Mary's is often a hallmark of damage to a *different* brain area—the ventromedial prefrontal cortex, or VMPFC, at the base of the frontal lobes. In contrast, DLPFC damage more commonly leads to an apathetic, disinterested state called abulia.

Why does it matter which part of the prefrontal cortex links to self-destructive behavior? Tumors can cause damage to both the VMPFC and the DLPFC, after all. Even if Lehrer glosses over the details, *some* part of the brain must be our rational center, right? Neuroethicists refer to this sort of simplified reasoning as "neurorealism"—the power of neuroscience data to make claims seem more real. Because he neatly maps our decision-making processes onto specific brain regions, Lehrer's subsequent advice seems grounded in our biology—even if he has interpreted the scientific facts to fit his story.

A second problem is that Lehrer misses areas of current and ongoing scientific debate. Consider the problem of delayed gratification: Many difficult real-world decisions, such as choosing where to invest money, require us to forgo an immediate reward to obtain a distant but better outcome. Lehrer discusses a seminal neuroimaging study in which participants chose to receive small gift certificates immediately or larger gift certificates in a few weeks. The results were striking: Nominally "emotional" brain regions became more active when rewards were available immediately, while nominally "rational" brain regions were equally activated regardless of delay.

Based on this study alone, one might conclude that the maladaptive influence of our brain's reward system causes impulsive choices. Several years ago, however, other neuroscientists ran a similar study and found a very different result: The activation of the brain's reward system signaled the subjective value of the gift certificate, *regardless of* the delay until its delivery. This second study suggested that there is no impulsive system to push people toward immediate rewards. This debate will remain unresolved until new research reconciles the conflicting results. What is clear, however, is that the neural basis of delayed gratification is more complex than Lehrer suggests.

Lehrer makes a similar error of omission when he discusses the risky behavior of teenagers. Noting that the prefrontal cortex is still immature during adolescence, Lehrer concludes that "teens make bad decisions because they are literally less rational." To anyone who has been around teenagers, this seems like a pretty safe statement—yet it may be wrong. Psychologist Valerie Reyna's work has shown that teenagers are, indeed, rational in their decision-making; that is, they weigh benefits, costs, and probabilities in a manner that fits standard economic models. Where they differ from adults, according to Reyna's research, is in their estimates about the consequences of their decisions. For instance, they underestimate the likelihood of getting pregnant following unsafe sex. These different explanations—inherent irrationality versus poor estimates of consequences—suggest very different strategies for preventing teenagers' risky behavior.

From Brain to Behavior: Practical Advice

Like many other popular neuroscience books, *How We Decide* ends with practical advice. To Lehrer's credit, he repeatedly states that no one rule for making decisions always works. Instead, the core challenge of decision-making lies in selecting a strategy that most likely will lead to an acceptable outcome. Some of these strategies can seem counterintuitive. For instance, recent studies suggest that we are more satisfied with our decisions if we distract ourselves beforehand. Here, Lehrer's recommendations frequently hit the mark.

Even so, one recurring piece of advice is troubling. Lehrer argues that because the prefrontal cortex is slow and has limited capacity, our rational faculties are poorly suited for complex decision-making: "We often make decisions on issues that are exceedingly complicated. In those situations, it's probably a mistake to consciously reflect on all the options, as this inundates the prefrontal cortex with too much data. ... When choosing a couch, or holding a mysterious set of cards, always listen to your feelings" (pages 236–237).

This is good advice for the expert poker player whose well-honed intuition incorporates hundreds of thousands of prior hands. But when we bring much less expertise to a problem—as when most of us balance our portfolios or play poker—our feelings may be a very poor guide indeed. Moreover, the learning abilities of our dopamine system, which Lehrer focuses on throughout the book, are inadequate for making many complicated decisions. Specifically, when we receive feedback infrequently, in abstract form, and without immediate personal consequence, our dopamine system is ill equipped to learn from our mistakes. For example, evaluating the terms of a mortgage without conscious reflection would be a bad idea, even when intuition screams that we should buy before housing prices rise.

How We Decide works best as a series of engaging anecdotes rather than a synthesis and interpretation of research. And contrary to the book jacket, this is not "the first book to use the unexpected discoveries of neuroscience to help us make the best decisions"—see earlier popular neuroscience books by Jason Zweig and Read Montague. Recent books by behavioral economists also provide similar sorts of advice, albeit with minimal neuroscience content.

Lehrer is a gifted storyteller, but too often he falls prey to neurorealism by providing a snapshot of neuroscience data to make a phenomenon seem more vivid and compelling. Even accurate summaries may omit the most exciting parts of cutting-edge research. For example, Lehrer points out that one standard drug treatment leads to pathological gambling in some Parkinson's patients, but he does not address the fascinating research question of what makes most patients resistant to the lure of gambling. Separately he describes the brain's insular cortex as activated by potential losses, whereas many scientists now think of this brain area like a switch: It shapes other regions' activity based upon the risk in the immediate environment.

Stories such as those in *How We Decide* reassure us. They reinforce our intuition and offer simple guidance. But the real world, in all its messy complexity, remains a far more interesting place to explore.

15

Synesthesia

Another World of Perception

Wednesday is Indigo Blue: Discovering the Brain of Synesthesia
By Richard E. Cytowic, M.D., and David M. Eagleman, Ph.D.
(The MIT Press, 2009; 320 pages, $29.95)

Reviewed by Julian E. Asher, Ph.D.

Julian E. Asher, Ph.D., is a research associate in the section of genomic medicine at Imperial College London and research director of the Synesthesia Research Group in the Department of Psychiatry at the University of Cambridge, which investigates the genetics of synesthesia.

SYNESTHESIA—from the Greek *syn* (union) and *aisthaesis* (sensation)—is a hereditary neurological condition in which ordinary activities trigger extraordinary experiences. A stimulus, such as a sound or a printed letter, produces a perception, such as color or taste. This condition has captured the popular imagination and has received considerable media attention, most recently following the pioneering identification of the first genetic regions linked to synesthesia reported earlier this year.[1] In their new book, *Wednesday Is Indigo Blue*, neurologist Richard E. Cytowic—one of the founders of modern synesthesia research—and neuroscientist David M. Eagleman provide a fascinating introduction to synesthesia, synesthetes (people who have the condition) and research on this condition.

The book's easy-to-digest overview of the growth of the field from relative obscurity to its current popularity enables people who are not familiar with synesthesia to dive in with relative ease. In addition to a thorough description and discussion of well-known forms of synesthesia, such as sound-color and grapheme-color (in which reading black text triggers the perception of color), the book contains the most cogent discussion to date of spatial sequence synesthesia (in which a person experiences sequences such as the days of the week as precisely ordered three-dimensional forms). This previously ignored variant is Eagleman's specialty, and his enthusiasm is thoroughly engaging. His pioneering development of a virtual reality environment where synesthetes can place the forms they perceive in space around a computer-generated avatar has led to a dramatic increase in our understanding of this form of synesthesia, which experts now acknowledge to be among the most common.

The book's discussion of taste synesthesia (encompassing taste both as a trigger and as a synesthetic response) is also particularly insightful, as we might expect given that taste-touch synesthesia was Cytowic's entry into the field. After discussing that initial case study at considerable length, Cytowic deftly integrates his and others' early explorations with more recent advances. Taste synesthesia is rare, but it has become an important research focus due to its implications for how synesthesia develops. For example, the tastes that synesthete James Wannerton experiences when he reads or hears words—the word *maybe* elicits the taste of baked beans, *most* tastes like toast and *department* tastes like jam tart—are based on Wannerton's childhood diet. This implies that these synesthetic associations were cemented during that period of his life.

Because perception lies at the heart of synesthesia, the authors of any book on the subject must find a way to convey synesthetes' perceptions to their primarily non-synesthetic audience. As neither author of *Wednesday Is Indigo Blue* is synesthetic, Cytowic and Eagleman wisely allow synesthetes to speak for themselves through quotations and anecdotes. While often less polished than the scientific portions of the text, these sections about real-life experiences offer a window into how synesthetes perceive the world. A sound-color synesthete describes her perception of her husband's voice as "a wonderful golden brown, like crisp, buttery toast," while a sound-taste synesthete describes the voices of people she knows as "barbecued pancakes," "spaghetti with M&M'S" and "peanut butter."

Most synesthetic perceptions are visual, and the frequent use of illustrations to show the reader what synesthetes "see" confirms that a picture really does speak a thousand words. This is especially the case when illustrating forms of synesthesia such as sequences in space, which are more difficult for readers to envision than are descriptions of synesthetes seeing a color when hearing a sound. The authors could have gone a step further, however; although the synesthetes' hand-drawn sketches are charming, the use of computer graphics to render some of them would have enabled the authors to illustrate visually salient texture and three-dimensional structure as well as color.

Cytowic and Eagleman provide the most balanced discussion thus far of how synesthesia affects the lives of synesthetes. Most of us assume that the condition dominates synesthetes' lives, but while their worlds may have a "different texture of reality," according to the authors, most synesthetes do not orient their lives around their synesthesia any more than most people orient their lives purely around their ability to see. While the authors occasionally fall into the trap of describing synesthesia as an "astonishing gift," they explicitly acknowledge that synesthesia is neither a gift nor a curse. This is particularly notable in light of the unfortunate tendency of some researchers and the popular media to overemphasize the condition's positive aspects (such as grapheme-color synesthetes' improved memory for names and numbers) while minimizing the very real disadvantages, such as sensory overload and cognitive difficulties. This popular bias distorts the public perception of synesthesia and does a tremendous disservice to synesthetes who find living with synesthesia challenging. It's not hard to imagine that seeing colors when you hear speech would make it difficult to follow the content of a lecture, or that math could trouble young synesthetes who haven't learned to separate synesthetic and symbolic meanings and thus find it perfectly natural to add a blue 3 and a yellow 2 to make a green 7. Eagleman and Cytowic also include anecdotes from several synesthetes who struggle with sensory overload. One woman describes her experience in London's Piccadilly Circus thus: "Every one of my senses is being battered. … It's like having nails at the back of my throat."

The authors provide similar balance in their discussion of the debate about whether synesthetes are more creative than non-synesthetes. The authors rightly point out that although people often assume that synes-thetes are more artistic, this is an artefact of sampling bias. Such claims stem from studies focusing on a few high-profile individuals such as composer Franz Liszt and artist David Hockney and ignoring the vast majority of non-artistic synesthetes. The confusion stems in large part from the use of the term *synesthesia* to describe an early 20th-century artistic movement whose followers attempted to meld visual art and music. There is some evidence that synesthetes are more creative in the broader sense of the word; for example, they score higher than non-synes-thetes on measures of flexibility and originality. However, creative doesn't necessarily equal artistic, and creative people can be found in any profes-sion in which thinking outside the box is helpful.

Cytowic and Eagleman do a superb job of situating synesthesia and its research within the larger context of cognitive neuroscience. They note that although there is something very different about the brains of synes-thetes, in many ways it is the similarities to non-synesthetes' brains that are most fascinating. Substantial evidence indicates that all humans are born synesthetic.[2, 3] While only some people retain explicit awareness of synesthetic perception, there are fundamental patterns (called form constants) underlying both synesthetic and non-synesthetic perception that will be readily apparent to any reader—for example, high musical tones are smaller and brighter while low tones are larger and darker. In addition to offering insight into the workings of the human brain, this discussion moves the book from being about "them" (synesthetes) to being about "us" (all humans).

While *Wednesday Is Indigo Blue* is an excellent introduction to the field, it is not without flaws. Discussion of the genetics of synesthesia is rather thin; it is limited to only a few pages focusing on a single paper about inheritance patterns. While this is an interesting introduction to heritability, the omission of the published molecular genetics work in this area is disappointing, particularly given the discussion of a putative "synesthesia gene"; the only mention of molecular genetic studies is a reference to preliminary results from Eagleman's research. Timing no doubt made it difficult to include the results, published this year,[1] of the first genetic study identifying candidate regions for synesthesia, but a discussion of the work done by other groups in this area and the existing molecular genetic studies on identical twins[4, 5] (which have important implications for Cytowic and Eagleman's discussion of inheritance patterns) would have added substance to the text.

Synesthesia research is a young and dynamic field with a number of active and interesting controversies. Unfortunately, the authors gloss over the central debate about the prevalence of synesthesia—which is, at its heart, a debate about the most fundamental question of all: What is synesthesia? Not all researchers agree about which phenomena should be considered forms of synesthesia, and prevalence figures vary widely depending on the definition used. The "broad tent" advocates argue for the inclusion of virtually any cross-modal phenomenon, whereas the "separate tent" advocates push for a more rigorous definition in the interest of clarity and preventing synesthesia from becoming a meaningless, catch-all term.

The authors' main flaw lies in treating the work of Julia Simner and colleagues on the prevalence of synesthesia[6] as definitive. Simner's estimate includes phenomena such as associations between the "notion of a person" and color, but these associations are not universally accepted as forms of synesthesia. Ironically, Simner and her colleagues explicitly acknowledge the uncertainty surrounding the definition of synesthesia, and Cytowic and Eagleman themselves express doubt about some forms of "personification" synesthesia. For example, is cross-sensory perception (such as combining sound and color) really part of the same phenomenon as cross-conceptual connections (such as linking numerical sequences to three-dimensional space) or the personification of inanimate objects? Would it make more sense from cognitive and neurobiological perspectives to classify some of these conditions as "related cross-modal phenomena" rather than as types of synesthesia? This is one of the key debates in the field, and the authors' failure to engage explicitly with it is disappointing.

Moreover, the authors' focus on their personal journeys through the field and their (considerable) contributions to it sometimes results in a less than balanced historical perspective. In particular, the contributions of Simon Baron-Cohen of the University of Cambridge (the other founder of the field) are given short shrift; Baron-Cohen's development of the test of "genuineness" to diagnose synesthesia[7] forms the foundation of modern research in this field and deserves more than a vague one-line mention.

On a structural note, distracting references to chapter 9 ("as we will see in chapter 9" and "which we will return to in chapter 9," for example) raises the question of whether the authors should have integrated that material into other parts of the book. It is probably the book's densest chapter, focusing primarily on the science rather than the experience of synesthesia, which might explain why it was left until the end. However, it's also one of the most interesting, with a discussion of synesthesia's potential neurological underpinnings. Integrating this material into the earlier chapters would have strengthened them and the work as a whole.

Despite its flaws, *Wednesday Is Indigo Blue* is an entertaining and informative introduction to synesthesia. It offers a window into a vastly different way of experiencing the world—and, more important, provides insight into what synesthetes' experiences can tell us about the human brain. It also casts light on the subjectivity of what we consider reality and reminds us that everyone experiences the world in a slightly different way; synesthesia is merely an extreme on that spectrum. Reality, as the authors say, is not one size fits all—and that's not a bad thing.

16

Weighing In on "Conditioned Hypereating"

The End of Overeating:
Taking Control of the Insatiable American Appetite
By David A. Kessler, M.D.
(Rodale Books, 2009; 320 pages, $25.95)

Reviewed by Lisa J. Merlo, Ph.D.,
and Mark S. Gold, M.D.

Lisa J. Merlo, Ph.D., is a licensed clinical psychologist, chief of the division of undergraduate education and director of the Addiction Medicine Public Health Research Group in the University of Florida Department of Psychiatry. She investigates psychosocial factors associated with risk, resiliency and treatment outcome in people with compulsive and addictive behavior disorders.

Mark S. Gold, M.D., is a Dizney Eminent Scholar and Distinguished Professor at the University of Florida College of Medicine's Brain Institute and chairman of the department of psychiatry. He has developed hypotheses and models for understanding the effects of drugs on the brain and behavior that have led to new treatment approaches for addiction.

"[That] is a load of fat on fat on fat and sugar that's then layered with fat on sugar on sugar and served with fat, salt, and fat... ." *(pp. 86–87)*

AS UNAPPETIZING AS IT SOUNDS, this is author David A. Kessler's striking way of describing a special dish at a popular chain restaurant, where sugar, fat and salt are purportedly combined in exacting proportions to create what one food industry executive calls "craveability." In *The End of Overeating: Taking Control of the Insatiable American Appetite*, Kessler further quotes this executive to illustrate how companies throughout the food industry engineer their products "to get you hooked" (p. 125). Kessler, former commissioner of the U.S. Food and Drug Administration, used to lead the fight against Big Tobacco. Now he has turned his attention to Big Food.

Although it remains unclear to what extent food industry executives understand the physiological and neurobiological impact of highly processed foods on consumers, it is obvious that they have uncovered the formula for creating mass-market products that stimulate repeated consumption. In his book, Kessler explores how the omnipresence of "hyperpalatable" foods in American society has given them increasing salience in our lives. It is difficult for many of us to ignore the siren call of the doughnut tray in the break room, the co-worker's candy dish, the drive-through restaurants on the way home or the bowl of ice cream we enjoy while watching the nightly news. Kessler argues that this "conditioned hypereating," the drive to eat beyond our needs, contributes to the global obesity epidemic and its innumerable consequences for our health.

The End of Overeating follows Kessler's quest to understand why so many people—including Kessler himself—lose control in the presence of food. In describing his plight, Kessler compares overeating to both pathological gambling and substance abuse. He writes that "it becomes an automatic response to widely available food and its cues, [and it is] excessive, driven by motivational forces we find difficult to control" (p. 145). Research supports Kessler's argument: The discovery that drugs assert their control over the brain by "hijacking" the neurobiological reward circuitry, rather than by merely producing withdrawal states, expanded the pool of potential substances of abuse to include food. Kessler suggests that because drugs, sex and food all affect the level of the neurotransmitter dopamine in the nucleus accumbens (the reward center of the brain), we may perceive foods that are hyperpalatable as more rewarding than those that are nutritious, thus increasing the likelihood that we will abuse the former.

The book is divided into six sections; the first three introduce concepts related to food production and marketing and how these processes influence our brains. Kessler describes how foods can become highly reinforcing stimuli as a result of neurobiological reward processes and environmental and social cues. For example, he presents the concept of "neuronal encoding" for palatability, where brain cells respond to rewarding foods and release electrochemical signals to stimulate other brain cells, and illustrates it with studies demonstrating that an animal will work almost as hard for a high-sugar, high-fat food as it will for cocaine. He discusses orosensory self-stimulation, which occurs when the ingestion of highly palatable foods stimulates the brain's natural opioid receptors, thereby encouraging the brain to crave more of these foods. In addition, Kessler reviews research that has shown how conditioning can increase cravings for certain foods.

This material prepares the reader for sections four and five, which discuss the theory and framework of "food rehab," strategies to curb addictive appetite and restore healthy eating habits. Though this book is not intended to serve as a diet manual, it does offer useful suggestions for controlling eating habits.

In the final section of the book, Kessler argues that policy changes requiring the food industry to adopt more-responsible practices could better safeguard the public from reinforced overeating. Kessler contends that improved regulation might help people recover from conditioned hypereating in the United States, much as tobacco regulations have curtailed smoking.

A quick read at 320 pages, this provocative volume stimulates the reader's desire for answers about the allure of hyperpalatable foods. The author's personal experiences and those of his acquaintances add color and context to Kessler's theory, and the concepts remain clear and relevant. Kessler supports his anecdotes with applicable research not only from brain-related disciplines such as psychology, neuroscience and psychiatry, but also from the fields of nutrition, medicine, culinary arts and marketing. He quotes material from personal interviews with the researchers in addition to reviewing and referencing scientific journal articles and textbooks.

In addition, Kessler focuses on broad concepts such as conditioning, reward, craving and addiction instead of following obscure, late-breaking discoveries involving specific genes, hormones or neurotransmitters. This approach makes the information more accessible to readers outside the field of neuroscience and prevents the book from becoming out-of-date almost immediately. Finally, Kessler provides 52 pages of endnotes that include both additional details and references for the research studies and interviews highlighted in the text. The comprehensive index also adds value; it lists material in the endnotes as well as in the main text.

Kessler could not cover everything in a relatively short volume, and some omissions will be particularly apparent to readers interested in neuroscience. For example, there is no discussion of the cutting-edge work of Kelly Brownell and his colleagues at the Rudd Center for Food Policy and Obesity at Yale University, who have examined how public health interventions, such as junk food taxes and nutrition labeling, affect the spread of obesity.[1] Second, Kessler introduces the term *conditioned hypereating* but does not refer to literature that explores the related concept of food addiction. First introduced in 1956,[2] food addiction has been covered rather extensively in the past decade.

Kessler also could have included elements of research that apply to his theory of conditioned hypereating, such as those covered in a special issue of the *Journal of Addiction Medicine* that I (Gold) edited earlier this year.[3] For example, Kessler seems not to have recognized the support for his theory in the accumulating imaging results, which Gene-Jack Wang of the Brookhaven National Laboratory published in a review describing how neuroimaging of brain dopamine pathways relates to obesity.[4] Moreover, Kessler should have discussed not just in the endnotes but in the text of his book the innovative animal studies by Bart Hoebel and colleagues at Princeton University,[5] who examined sugar addiction in rats. Next, Kessler ought to have included recent work implicating high-fructose corn syrup in the many negative health effects of obesity and overeating. For example, reference to the work of Richard J. Johnson of the University of Colorado Health Sciences Center, who wrote *The Sugar Fix* in 2008, would have strengthened Kessler's book.

Finally, given Kessler's heavy reliance on primary sources' personal experiences, it is curious that he did not include at least one interview with someone who had undergone bariatric surgery or lap-band surgery. Because they often experience a drastic shift in their relationship with food, these individuals would have provided a unique perspective on Kessler's theory.

But these omissions do not detract from the fascinating nature of the book or its potential impact on public health. On the whole, *The End of Overeating* is thought-provoking and soundly based in science. It also provides some preliminary suggestions for conquering conditioned hypereating. Be forewarned: After digesting this volume, you might not be able to look at food in the same way. You might find yourself critiquing the descriptions of entrées on restaurant menus, studying food labels and discarding some of your former favorite foods. Kessler would approve of such behaviors because they would contribute to a "critical perceptual shift" (p. 234) in public opinion toward food. This shift will be necessary for American society to overcome the epidemic of conditioned hypereating.

17

Our Neurotech Future

The Neuro Revolution:
How Brain Science Is Changing Our World
By Zack Lynch with Byron Laursen
(St. Martin's Press, 2009; 256 pages, $25.99)

Reviewed by Michael F. Huerta, Ph.D.*

Michael F. Huerta, Ph.D., is the associate director for scientific technology research at the National Institute of Mental Health and leads neuroscience and neurotechnology programs and initiatives at the National Institutes of Health (NIH). He directs the NIH's Human Connectome Project, the National Database for Autism Research and the Office of Cross-Cutting Science and co-chairs the coordinating committee of the NIH Blueprint for Neuroscience Research.

---

* The views expressed in this book review do not necessarily represent the views of the National Institute of Mental Health, the National Institutes of Health, the U.S. Department of Health and Human Services, or the United States government.

TECHNOLOGY'S POTENTIAL to improve—or to imperil—our lives and our societies lies at the center of this entertaining and thought-provoking book by Zack Lynch, founder and executive director of the Neurotechnology Industry Organization.

Written for a lay audience, *The Neuro Revolution* begins with Lynch's description of his first bungee jump, from the canopy of a lush rain forest, followed by a shock of pain from an injured spine when his second jump went awry. The experience inspired him to explore neuroscience and neurotechnology (Lynch defines the latter as "the tools we use to understand and influence our brain and nervous system"). With prose that is at times clever and quirky but never dull, Lynch discusses how our understanding of the human brain—as well as our ability to influence it—may shape the future of law, commerce, art, warfare and religion.

Along the way, we read stories of discovery and invention set in a variety of contexts and disciplines. Lynch's anecdotes illustrate how the findings and technologies of brain science might alter society. He supports his stories and personal musings with references to reader-friendly articles and books; comments and insights from scientists, artists, ethicists and other experts; and historical facts that help the reader appreciate the full trajectory of a discovery.

In addition, Lynch draws heavily on his market-oriented perspective—the Neurotechnology Industry Organization represents companies involved in neuroscience and brain research, as well as patient advocacy groups, and Lynch co-founded a market research firm that focuses on the impact of neurotechnology on business, government and society. Business interests aside, his sketching of the potential commercial significance of new, brain-relevant technologies makes his predictions more compelling. For example, he cites the use of magnetic resonance imaging or detection of electric signals to monitor one's brain activity and relate it to one's behavior, which would have commercial value for purposes as wide-ranging as anticipating a person's response to a drug or to an advertisement. In a world driven by market forces, the marketplace will strongly influence how our future, both as individuals and as a society, will unfold.

Each chapter of *The Neuro Revolution* focuses on one way that neuroscience will influence our future. The societal ramifications of each chapter's content, however, are broad, reflecting the expansive nature of the technologies. For instance, the same technologies that Lynch describes in the chapter about legal implications beyond research and standard medical practice might also be relevant to chapters concerning war or commerce. As an example, the use of neuroimaging data and their computational analyses to assess the veracity of witness testimony might also apply to their use in interrogating enemy soldiers or identifying products that are particularly appealing to an individual. As the title of the book indicates, Lynch makes clear that the impact neurotechnological discoveries might have on our world compels us to consider these innovations from a societal, as well as a scientific, perspective.

Readers already familiar with neurotechnology will recognize many of Lynch's topics. For example, Lynch discusses the past, present and future influences of lie-detection technology in the courtroom, an increasingly prominent subject as the idea of using brain imaging to detect lies gains traction. He points out that several federal national security agencies have applied neuroscience and related technology to try to detect lies, However, Lynch notes that sophisticated technology is not necessary reliable in every possible application, and we must be careful not to accept claims that scientists have yet to verify. In the courtroom, the consequences of trusting in unproven technology can be significant.

To provide background on the use of lie-detection tools in the legal system, Lynch offers stories and examples based in legal fact. Then he details how neurological studies of memory might be useful in developing the next generation of lie detectors. Lynch describes this and other neuro-technologies with characteristic zeal and animation without wandering too far from rigorous scientific interpretation. This is a difficult balance to achieve.

Ethicists and scientists are becoming increasingly interested in using neurotechnology to enhance—or, as Lynch prefers, to "enable"—human functioning. Currently, physical and mental "enhancement" is possible primarily through drugs, but as researchers find new and better ways to link the brain with computer technology, neurological devices meant to improve performance will become more common. The notion of using technology, pharmacologic or otherwise, for this purpose is not new. Professional athletes have been using steroids for years, and the U.S. military has long provided stimulants to pilots flying long-range missions. But with increased specificity of drugs and the development of increasingly sophisticated devices, such as neural prostheses, products that maximize brain function have become a growth market. Lynch points out that as new technologies allow us to modify different brain functions in different ways—for example, enhancing cognition while simultaneously suppressing emotion—we must consider how these capabilities affect the human identity.

Near the end of the book, Lynch asserts that during the next 30 years neuroscience and neurotechnology will produce a "neurosociety" in which "you will eventually be able to continuously shape your emotional stability, sharpen your mental clarity, and extend your most desirable sensory states until they become your dominant experience of reality." This prediction, which is loaded with existential implications, might seem overly bold or even fanciful. But to anyone familiar with neurotechnology, and to anyone finishing *The Neuro Revolution*, this view of the future may seem reasonable or even conservative.

Given the broad perspective Lynch takes in this book, readers may argue that he should have addressed additional technologies, neuroscience findings and ethical or legal implications more thoroughly. However, suggesting that additional points could be made or that topics could be explored further is more an acknowledgement of the richness and import of Lynch's subject matter than a criticism of what this slim book offers.

As we continue to learn more about our underlying biology, and as it becomes easier to modify our bodies, momentous questions are arising. For example, when drugs or devices are used to augment brain function in healthy individuals rather than to treat a disorder, where should we set personal, ethical, legal and medical thresholds? People from all walks of life will soon need to address these kinds of questions. *The Neuro Revolution* is a timely and approachable introduction to the power of neuroscience and neurotechnology to shape our world, inside and out.

Endnotes

1. THE SCIENCE OF EDUCATION: INFORMING TEACHING AND LEARNING THROUGH THE BRAIN SCIENCES

1. P. Shaw, K. Eckstrand, W. Sharp, J. Blumenthal, J. P. Lerch, D. Greenstein, L. Clasen, A. Evans, J. Giedd, and J. L. Rapoport, "Attention-deficit/ Hyperactivity Disorder Is Characterized by a Delay in Cortical Maturation," *Proceedings of the National Academy of Sciences* 104 (2007): 19649–19654.

2. M. H. Immordino-Yang and A. Damasio, "We Feel, Therefore We Learn: The Relevance of Affective and Social Neuroscience to Education," *Mind, Brain, and Education* 1, no. 1 (2007): 3–10.

3. J. Fan, J. I. Flombaum, B. D. McCandliss, K. M. Thomas, and M. I. Posner, "Cognitive and Brain Consequences of Conflict," *Neuro Image* 18 (2003): 42–57.

4. H. Gardner, "Quandaries for Neuroeducation," *Mind, Brain, and Education* 2, no. 4 (2008): 165–169.

5. M. M. Mazzocco, "Introduction: Mathematics Ability, Performance, and Achievement," *Developmental Neuropsychology* 33, no. 3 (2008): 197–204.

6. M. Brabeck, "Why We Need 'Translational' Research: Putting Clinical Findings to Work in Classrooms," *Education Week* 27, no. 38 (2008): 28, 36.

7. K. L. Hyde, J. Lerch, A. Norton, M. Forgeard, E. Winner, A. C. Evans, and G. Schlaug, "Musical Training Shapes Structural Brain Development," *Journal of Neuroscience* 29 (2009): 3019–3025.

8. B. Wandell, R. Dougherty, M. Ben-Shachar, G. Deutsch, and J. Tsang, "Training in the Arts, Reading, and Brain Imaging," *Learning, Arts, and the Brain: The Dana Consortium Report* (2008): 51–59.

9. P. I. Yakovlev and A. R. Lecours, "The Myelogenetic Cycles of Regional Maturation of the Brain," in *Regional Development of the Brain in Early Life*, ed. A. Minkowsky, 3–70 (Oxford, England: Blackwell Scientific, 1967).

2. HOW ARTS TRAINING IMPROVES ATTENTION AND COGNITION

1. M. I. Posner and M. K. Rothbart, "Research on Attention Networks as a Model for the Integration of Psychological Science," *Annual Review of Psychology* 58 (2007): 1–23.

2. F. H. Rauscher, G. L. Shaw, and C. N. Ky, "Music and Spatial Task Performance," *Nature* 365 (1993): 611.

3. E. G. Schellenberg, "Music Lessons Enhance IQ," *Psychological Science* 15 (2004): 511–514.

4. K. L. Hyde, J. Lerch, A. Norton, M. Forgeard, E. Winner, A. C. Evans, and G. Schlaug, "Musical Training Shapes Structural Brain Development," *Journal of Neuroscience* 29 (2009): 3019–3025.

5. M. R. Rueda, M. I. Posner, and M. K. Rothbart, "Attentional Control and Self Regulation" in *Handbook of Self Regulation: Research, Theory, and Applications*, ed. R. F. Baumeister and K. D. Vohs, 283–300 (New York: Guilford Press, 2004).

6. P. Checa, R. Rodriguez-Bailon, and M. R. Rueda, "Neurocognitive and Temperamental Systems of Early Self-Regulation and Early Adolescents' Social and Academic Outcomes," *Mind Brain and Education* 2 (2008): 177–187.

7. M. R. Rueda, M. K. Rothbart, B. D. McCandliss, L. Saccomanno, and M. I. Posner, "Training, Maturation and Genetic Influences on the Development of Executive Attention," *Proceedings of the National Academy of Sciences* 102 (2005): 4931–4936.

8. J. Fan, J.I. Flombaum, B.D. McCandliss, K.M. Thomas, and M.I. Posner, "Cognitive and Brain Consequences of Conflict," *Neuro Image* 18 (2003): 42–57.

9. A. Diamond, S. Barnett, J. Thomas, and S. Munro, "Preschool Program Improves Cognitive Control," *Science* 318 (2007): 1387–1388.

10. E. Sheese, M. Pascale, M. Voelker, M. K. Rothbart, and M. I. Posner, "Parenting Quality Interacts with Genetic Variation in Dopamine Receptor D4 to Influence Temperament in Early Childhood," *Development and Psychopathology* 19, no. 4 (2007): 1039–1046.

11. G. A. Bryant and H. C. Barrett, "Recognizing Intentions in Infant-directed Speech: Evidence for Universals," *Psychological Science* 18, no. 8 (2007): 746–751.

12. B. Wandell, R. Dougherty, M. Ben-Shachar, G. Deutsch, and J. Tsang, "Training in the Arts, Reading, and Brain Imaging," *Learning, Arts, and the Brain: The Dana Consortium Report* 51-59.

13. E. Spelke, "Effects of Music Instruction on Developing Cognitive Systems at the Foundations of Math and Science," *Learning, Arts, and the Brain: The Dana Consortium Report* 17-49.

14. Y. Tang, Y. Ma, Y. Fan, H. Feng, J. Wang, S. Feng, Q. Lu, B. Hu, Y. Lin, J. Li, Y. Zhang, Y. Wang, L. Zhou, and M. Fan, "Central and Autonomic Nervous System Interaction is Altered by Short Term Meditation, *Proceedings of the National Academy of Science* 106(2009): 8865–8870.

3. WHAT CAN DANCE TEACH US ABOUT LEARNING?

1. R. A. Schmidt and T. D. Lee, *Motor Control and Learning* (Champaign, IL: Human Kinetics, 1999).

2. S. T. Grafton, "Embodied Cognition and the Simulation of Action to Understand Others," *Annals of the New York Academy of Sciences* 1156 (2009): 97–117.

3. B. Calvo-Merino, D. E. Glaser, J. Grèzes, R. E. Passingham, and P. Haggard, "Action Observation and Acquired Motor Skills: an fMRI Study with Expert Dancers," *Cerebral Cortex* 15, no. 8 (2005): 1243–1249.

4. B. Calvo-Merino, J. Grèzes, D. E. Glaser, R. E. Passingham, and P. Haggard, "Seeing or Doing? Influence of Visual and Motor Familiarity in Action Observation," *Current Biology* 16 (2006): 1905–1910.

5. S. M. Aglioti, P. Cesari, M. Romani, and C. Urgesi, "Action Anticipation and Motor Resonance in Elite Basketball Players," *Nature Neuroscience* 11 (2008): 1109–1116.

6. E. S. Cross, A. F. Hamilton, and S. T. Grafton, "Building a Motor Simulation De Novo: Observation of Dance by Dancers," *Neuroimage* 31, no. 3 (2006): 1257–1267.

7. E. S. Cross, D. J. Kraemer, A. F. Hamilton, W. M. Kelley, and S. T. Grafton, "Sensitivity of the Action Observation Network to Physical and Observational Learning," *Cerebral Cortex* 19, no. 2 (2009): 315–326.

8. E. S. Cross, A. F. Hamilton, D. J. M. Kraemer, W. M. Kelley, and S. T. Grafton. "Dissociable Substrates for Body Motion and Physical Experience in the Human Action Observation Network," *European Journal of Neuroscience*, in press.

9. E. G. Schellenberg, "Music Lessons Enhance IQ," *Psychological Science* 15, no. 8 (2004): 511–514.

10. Y. Y. Tang and M. I. Posner, "Attention Training and Attention State Training," *Trends in Cognitive Sciences* 13, no. 5 (2009): 222–227.

11. M. B. Crawford, *Shop Class as Soulcraft: An Inquiry into the Value of Work* (New York: Penguin Press, 2009).

12. S. Ortigue, F. Bianchi-Demicheli, A. F. Hamilton, and S. T. Grafton, "The Neural Basis of Love as a Subliminal Prime: An Event-related Functional Magnetic Resonance Imaging Study," *Journal of Cognitive Neuroscience* 19, no. 7 (2007): 1218–1230.

13. K. Adolph, "Learning to Move," *Psychological Science* 17, no. 3 (2008): 213–218.

5. THE TEEN BRAIN:
PRIMED TO LEARN, PRIMED TO TAKE RISKS

1. L. P. Spear, "The Adolescent Brain and Age-Related Behavioral Manifestations," *Neuroscience and Biobehavioral Reviews* 24, no. 4 (2000): 417.

2. Centers for Disease Control and Prevention Health Data Interactive, "Mortality by Underlying and Multiple Cause, Ages 18+: US, 1981–2005 (Source: NVSS)," http://205.207.175.93/hdi/ReportFolders/ReportFolders.aspx?IF_ActivePath=P,21 (accessed February 23, 2009).

3. F. M. Benes, in C. A. Nelson and M. Luciana, eds., *Handbook of Developmental Cognitive Neuroscience* (Cambridge, MA: MIT Press, 2001), 79.

4. R. D. Fields, "White Matter in Learning, Cognition, and Psychiatric Disorders," *Trends in Neurosciences* 31, no. 7 (2008): 361.

5. Z. Nagy, H. Westerberg, and T. Klingberg, "Maturation of White Matter Is Associated with the Development of Cognitive Functions during Childhood," *Journal of Cognitive Neuroscience* 16, no. 7 (2004): 1227.

6. G. K. Deutsch, R. F. Dougherty, R. Bammer, W. T. Siok, J. D. Gabrieli, and B. Wandell, "Children's Reading Performance Is Correlated with White Matter Structure Measured by Diffusion Tensor Imaging," *Cortex* 41, no. 3 (2005): 354.

7. C. Liston, R. Watts, N. Tottenham, M. C. Davidson, S. Niogi, A. M. Ulug, and B. J. Casey, "Frontostriatal Microstructure Modulates Efficient Recruitment of Cognitive Control," *Cerebral Cortex* 16, no. 4 (2006): 553.

8. V. Menon and S. Crottaz-Herbette, "Combined EEG and fMRI Studies of Human Brain Function," *International Review of Neurobiology* 66 (2005): 291.

9. M. H. Grosbras, M. Jansen, G. Leonard, A. McIntosh, K. Osswald, C. Poulsen, L. Steinberg, R. Toro, and T. Paus, "Neural Mechanisms of Resistance to Peer Influence in Early Adolescence," *Journal of Neuroscience* 27, no. 30 (2007): 8040.

10. N. Gogtay, J. N. Giedd, L. Lusk, K. M. Hayashi, D. Greenstein, A. C. Vaituzis, T. F. Nugent III, D. H. Herman, L. S. Clasen, A. W. Toga, J. L. Rapoport, and P. M. Thompson, "Dynamic Mapping of Human Cortical Development during Childhood through Early Adulthood," *Proceedings of the National Academy of Sciences of the United States of America* 101, no. 21 (2004): 8174.

11. J. M. Bjork, B. Knutson, G. W. Fong, D. M. Caggiano, S. M. Bennett, and D. W. Hommer, "Incentive-Elicited Brain Activation in Adolescents: Similarities and Differences from Young Adults," *Journal of Neuroscience* 24, no. 8 (2004): 1793.

6. VIDEO GAMES AFFECT THE BRAIN – FOR BETTER *AND* WORSE

1. D. A. Gentile and J. R. Gentile, "Violent Video Games as Exemplary Teachers: A Conceptual Analysis," *Journal of Youth and Adolescence* 9 (2008): 127–141.

2. R. F. Murphy, W. R. Penuel, B. Means, C. Korbak, A. Whaley, and J. E. Allen, *A Review of Recent Evidence on the Effectiveness of Discrete Educational Software* (Washington, DC: Planning and Evaluation Service, U.S. Department of Education, 2002).

3. C. S. Green and D. Bavelier, "Action Video Game Modifies Visual Selective Attention," *Nature* 423 (2003): 534–537.

4. R. Li, U. Polat, W. Makous, and D. Bavelier, "Enhancing the Contrast Sensitivity Function through Action Video Game Training," *Nature Neuroscience* 12 (2009): 549–555.

5. R. Hämäläinen, T. Manninen, S. Järvela, and P. Häkkinen, "Learning to Collaborate: Designing Collaboration in a 3-D Game Environment," *Internet and Higher Education* 9, no. 1 (2006): 47–61.

6. D. A. Gentile, C. A. Anderson, S. Yukawa, M. Saleem, K. M. Lim, A. Shibuya, A. K. Liau, A. Khoo, B. J. Bushman, L. R. Huesmann, and A. Sakamoto, "The Effects of Prosocial Video Games on Prosocial Behaviors: International Evidence from Correlational, Longitudinal, and Experimental Studies," *Personality and Social Psychology Bulletin* (2009).

7. J. C. Rosser Jr., P. J. Lynch, L. Cuddihy, D. A. Gentile, J. Klonsky, and R. Merrell, "The Impact of Video Games on Training Surgeons in the 21st Century," *Archives of Surgery* 142, no. 2 (2007): 181–186.

8. C. A. Anderson, "An Update on the Effects of Playing Violent Video Games," *Journal of Adolescence* 27 (2004): 113–122.

9. C. A. Anderson, A. Sakamoto, D. A. Gentile, N. Ihori, A. Shibuya, S. Yukawa, M. Naito, and K. Kobayashi, "Longitudinal Effects of Violent Video Games on Aggression in Japan and the United States," *Pediatrics* 122, no. 5 (2008): e1067–1072.

10. M. J. Koepp, R. N. Gunn, A. D. Lawrence, V. J. Cunningham, A. Dagher, T. Jones, D. J. Brooks, C. J. Bench, and P. M. Grasby, "Evidence for Striatal Dopamine Release during a Video Game," *Nature* 393 (1998): 266–268.

11. R. Weber, U. Ritterfeld, and K. Mathiak, "Does Playing Violent Video Games Induce Aggression? Empirical Evidence of a Functional Magnetic Resonance Imaging Study," *Media Psychology* 8, no. 1 (2006): 39–60.

12. B. D. Bartholow, B. J. Bushman, and M. A. Sestir, "Chronic Violent Video Game Exposure and Desensitization to Violence: Behavioral and Event-Related Brain Potential Data," *Journal of Experimental Social Psychology* 42, no. 4 (2006): 532–539.

13. D. Lieberman, "Management of Chronic Pediatric Diseases with Interactive Health Games: Theory and Research Findings," *Journal of Ambulatory Care Management* 24, no. 1 (2001): 26–38.

14. S. M. Jaeggi, M. Buschkuel, J. Jonides, and W. J. Perrig, "Improving Fluid Intelligence with Training on Working Memory," *Proceedings of the National Academy of Sciences* 105, no. 19 (2008): 6829–6833.

7. UPDATING THE *DIAGNOSTIC AND STATISTICAL MANUAL OF MENTAL DISORDERS*

Kupfer et al. References

1. J. P. Feighner, E. Robins, S. B. Guze, R. A. Woodruff Jr., G. Winokur, and R. Munoz, "Diagnostic Criteria for Use in Psychiatric Research," *Archives of General Psychiatry* 26, no. 1 (1972): 57–63.

2. E. Robins and S. B. Guze, "Establishment of Diagnostic Validity in Psychiatric Illness: Its Application to Schizophrenia," *American Journal of Psychiatry* 126, no. 7 (1970): 983–987.

3. R. L. Spitzer, J. Endicott, and E. Robins, "Research Diagnostic Criteria: Rationale and Reliability," *Archives of General Psychiatry* 35, no. 6 (1978): 773–782.

4. J. Endicott and R. L. Spitzer, "A Diagnostic Interview: The Schedule for Affective Disorders and Schizophrenia," *Archives of General Psychiatry* 35, no. 7 (1978): 837–844.

5. D. A. Regier, J. K. Myers, M. Kramer, L. N. Robins, D. G. Blazer, R. L. Hough, W. W. Eaton, and B. Z. Locke, "The NIMH Epidemiologic Catchment Area (ECA) Program: Historical Context, Major Objectives, and Study Population Characteristics," *Archives of General Psychiatry* 41, no. 10 (1984): 934–941.

6. J. H. Boyd, J. D. Burke Jr., E. Gruenberg, C. E. Holzer III, D. S. Rae, L. K. George, M. Karno, R. Stoltzman, L. McEvoy, and G. Nestadt, "Exclusion Criteria of DSM-III: A Study of Co-occurrence of Hierarchy-free Syndromes," *Archives of General Psychiatry* 41, no. 10 (1984): 983–989.

7. R. H. Howland, J. A. Rush, S. R. Wisniewski, M. H. Trivedi, D. Warden, M. Fava, L. L. Davis, G. K. Balasubramani, P. J. McGrath, and S. R. Berman, "Concurrent Anxiety and Substance Use Disorders among Outpatients with Major Depression: Clinical Features and Effect on Treatment Outcome," *Drug and Alcohol Dependence* 99, no. 1 (2009): 248–260.

8. B. Löwe, R. L. Spitzer, J. B. W. Williams, M. Mussell, D. Schellberg, and K. Kroenke, "Depression, Anxiety and Somatization in Primary Care: Syndrome Overlap and Functional Impairment," *General Hospital Psychiatry* 30, no. 3 (2008): 191–199.

9. R. L. Kendell and A. Jablensky, "Distinguishing between the Validity and Utility of Psychiatric Diagnoses," *American Journal of Psychiatry* 160, no. 1 (2003): 4–12.

10. D. A. Regier, W. E. Narrow, E. A. Kuhl, and D. J. Kupfer, "The Conceptual Development of DSM-V," *American Journal of Psychiatry* 166, no. 6 (2009): 645–650.

11. F. F. Duffy, H. Chung, M. Trivedi, D. S. Rae, D. A. Regier, and D. J. Katzelnick, "Systematic Use of Patient-Rated Depression Severity Monitoring: Is It Helpful and Feasible in Clinical Psychiatry?" *Psychiatric Services* 59, no. 10 (2008): 1148–1154.

12. N. Craddock, M. C. O'Donovan, and M. J. Owen, "Genes for Schizophrenia and Bipolar Disorder? Implications for Psychiatric Nosology," *Schizophrenia Bulletin* 23, no. 1 (2006): 9–16.

13. S. E. Hyman, "A Glimmer of Light for Neuropsychiatric Disorders," *Nature* 455, no. 7215 (2008): 890–893.

14. The International Schizophrenia Consortium, "Common Polygenic Variation Contributes to Risk of Schizophrenia and Bipolar Disorder," *Nature* 460, no. 7256 (2009): 748–752.

15. P. F. Sullivan, K. S. Kendler, and M. C. Neale, "Schizophrenia as a Complex Trait: Evidence from a Meta-analysis of Twin Studies," *Archives of General Psychiatry* 60, no. 12 (2003): 1187–1192.

16. G. Andrews, D. P. Goldberg, R. F. Krueger, W. T. Carpenter, S. E. Hyman, P. Sachdev, and D. S. Pine, "Exploring the Feasibility of a Meta-Structure for DSM-V and ICD-11," *Psychological Medicine*, in press.

McHugh References

1. R. L. Spitzer and J. L. Fleiss, "A Re-analysis of the Reliability of Psychiatric Diagnosis," *British Journal of Psychiatry* 125 (1974): 341–347.

2. A. V. Horwitz and J. Wakefield, *The Loss of Sadness: How Psychiatry Transformed Normal Sorrow into Depressive Disorder* (England: Oxford University Press, 2007).

McHugh General References

P. R. McHugh, "Striving for Coherence: Psychiatry's Efforts over Classification," *Journal of the American Medical Association* 293 (2005): 2526–2528.

P. R. McHugh and P. R. Slavney, *The Perspectives of Psychiatry*, 2nd ed. (Baltimore: Johns Hopkins Press, 1998).

8. USING DEEP BRAIN STIMULATION ON THE MIND: HANDLE WITH CARE

1. F. M. Weaver, K. Follett, M. Stern, K. Hur, C. Harris, W. J. Marks Jr., J. Rothlind, O. Sagher, D. Reda, C. S. Moy, R. Pahwa, K. Burchiel, P. Hogarth, E. C. Lai, J. E. Duda, K. Holloway, A. Samii, S. Horn, J. Bronstein, G. Stoner, J. Heemskerk, and G. D. Huang; CSP 468 Study Group, "Bilateral Deep Brain Stimulation vs Best Medical Therapy for Patients with Advanced Parkinson Disease: A Randomized Controlled Trial," *Journal of the American Medical Association* 301, no. 1 (2009): 63–73.

2. J. W. Mink, J. Walkup, K. A. Frey, P. Como, D. Cath, M. R. DeLong, G. Erenberg, J. Jankovic, J. Juncos, J. F. Leckman, N. Swerdlow, V. Visser-Vandewalle, and J. L. Vitek; Tourette Syndrome Association, Inc., "Patient Selection and Assessment Recommendations for Deep Brain Stimulation in Tourette Syndrome," *Movement Disorders* 21, no. 11 (2006): 1831–1838.

9. NEUROIMAGING:
SEPARATING THE PROMISE FROM THE PIPE DREAMS

1. "Frequently Asked Questions," FKF Applied Research Inc., http://fkfftp.com/ FAQS.html.

2. D. G. Amen, C. Hanks, and J. Prunella, "Predicting Positive and Negative Treatment Responses to Stimulants with Brain SPECT Imaging," *Journal of Psychoactive Drugs* 40, no. 2 (2008): 131–138.

3. Council on Children, Adolescents, and Their Families, *Brain Imaging and Child and Adolescent Psychiatry with Special Emphasis on Single Photon Emission Computed Tomography (SPECT)* (Arlington, VA: The American Psychiatric Association, 2005).

General References

H. T. Greely and J. Illes, "Neuroscience-Based Lie Detection: The Urgent Need for Regulation," *American Journal of Law and Medicine* 33, nos. 2–3 (2007): 377–431.

N. K. Logothetis, "What We Can Do and What We Cannot Do with fMRI," *Nature* 453, no. 7197 (2008): 869–878.

A. J. O'Toole, F. Jiang, H. Abdi, N. Penard, J. P. Dunlop, and M. A. Parent, "Theoretical, Statistical, and Practical Perspectives on Pattern-Based Classification Approaches to the Analysis of Functional Neuroimaging Data," *Journal of Cognitive Neuroscience* 19, no. 11 (2007): 1735–1752.

R. A. Poldrack, "Can Cognitive Processes Be Inferred from Neuroimaging Data?" *Trends in Cognitive Science* 10, no. 2 (2006): 59–63.

E. Racine, O. Bar-Ilan, and J. Illes, "fMRI in the Public Eye." *Nature Reviews Neuroscience* 6, no. 2 (2005): 159–64.

10. WHY SO MANY SENIORS GET SWINDLED:
BRAIN ANOMALIES AND POOR DECISION-MAKING IN OLDER ADULTS

1. R. West, "An Application of Prefrontal Cortex Function Theory of Cognitive Aging," *Psychological Bulletin* 120 (1996): 272.

2. A. R. Damasio, *Descartes' Error: Emotion, Reason, and the Human Brain* (New York: Grosset/Putnam, 1994), 165–204.

3. A. Bechara, "Decision-making, Impulse Control and Loss of Willpower to Resist Drugs: A Neurocognitive Perspective," *Nature Neuroscience* 8 (2005): 1458.

4. A. Bechara, D. Tranel, and H. Damasio, "Characterization of the Decision-Making Deficit of Patients with Ventromedial Prefrontal Cortex Lesions," *Brain* 123 (2000): 2189.

5. N. L. Denburg, C. A. Cole, M. Hernandez, T. H. Yamada, D. Tranel, A. Bechara, and R. B. Wallace, "The Orbitofrontal Cortex, Real-World Decision-making, and Aging," *Annals of the New York Academy of Sciences* 1121 (2007): 480.

6. R. D. Rogers, B. J. Everitt, A. Baldacchino, A. J. Blackshaw, R. Swainson, K. Wynne, N. B. Baker, J. Hunter, T. Carthy, E. Booker, M. London, J. F. Deakin, B. J. Sahakian, and T. W. Robbins, "Dissociable Deficits in the Decision-making Cognition of Chronic Amphetamine Abusers, Opiate Abusers, Patients with Focal Damage to Prefrontal Cortex, and Tryptophan-Depleted Normal Volunteers: Evidence for Monoaminergic Mechanisms," *Neuropsychopharmacology* 20 (1999): 322.

11. WIRED FOR HUNGER: THE BRAIN AND OBESITY

1. Y. Zhang, R. Proenca, M. Maffei, M. Barone, L. Leopold, and J. M. Friedman, "Positional Cloning of the Mouse Obese Gene and Its Human Homologue," *Nature* 372, no. 6505 (1994): 425–432.

2. A. M. Cali and S. Caprio, "Obesity in Children and Adolescents," *Journal of Clinical Endocrinological Metabolism* 93(11 Suppl 1, 2008): S31–S36.

3. C. Sherrington, "Cutaneous Sensation," in *Textbook of Physiology*, ed. E. A. Sharpey-Schaefer, 920–1001 (Edinburgh: Young J. Pentland, 1900).

4. D. L. Coleman and K. P. Hummel, "Effects of Parabiosis of Normal with Genetically Diabetic Mice," *American Journal of Physiology* 217 (1969): 1298–1304.

5. D. L. Coleman, "Effects of Parabiosis of Obese with Diabetes and Normal Mice," *Diabetologia* 9 (1973): 294–298.

6. D. L. Coleman, "Diabetes-Obesity Syndromes in Mice," *Diabetes* 31 (1982): 1–6.

7. N. R. Lenard and H. R. Berthoud, "Central and Peripheral Regulation of Food Intake and Physical Activity: Pathways and Genes," *Obesity* 16 (Suppl 3, 2008): S11–S22.

8. W. A. Banks, A. J. Kastin, W. Huang, J. B. Jaspan, and L. M. Maness, "Leptin Enters the Brain by a Saturable System Independent of Insulin," *Peptides* 17, no. 2 (1996): 305–311.

9. N. Satoh, Y. Ogawa, G. Katsuura, M. Hayase, T. Tsuji, K. Imagawa, Y. Yoshimasa, S. Nishi, K. Hosoda, and K. Nakao, "The Arcuate Nucleus as a Primary Site of Satiety Effect of Leptin in Rats," *Neuroscience Letters* 224, no. 3 (1997): 149–152.

10. S. Pinto, A. G. Roseberry, H. Liu, S. Diano, M. Shanabrough, X. Cai, J. M. Friedman, and T. L. Horvath, "Rapid Rewiring of Arcuate Nucleus Feeding Circuits by Leptin," *Science* 304 (2004): 110–115.

11. P. J. Scarpace and Y. Zhang, "Leptin Resistance: A Predisposing Factor for Diet-induced Obesity," *American Journal of Physiology - Regulatory, Integrative and Comparative Physiology* 296, no. 3 (2009): R493–R500.

12. M. Shintani, Y. Ogawa, K. Ebihara, M. Aizawa-Abe, F. Miyanaga, K. Takaya, T. Hayashi, G. Inoue, K. Hosoda, M. Kojima, K. Kangawa, and K. Nakao, "Ghrelin, an Endogenous Growth Hormone Secretagogue, Is a Novel Orexigenic Peptide That Antagonizes Leptin Action through the Activation of Hypothalamic Neuropeptide Y/Y1 Receptor Pathway," *Diabetes* 50, no. 2 (2001): 227–232.

13. C. B. Lawrence, A. C. Snape, F. M. Baudoin, and S. M. Luckman, "Acute Central Ghrelin and GH Secretagogues Induce Feeding and Activate Brain Appetite Centers," *Endocrinology* 143, no. 1 (2002): 155–162.

14. L. Wang, D. H. Saint-Pierre, and Y. Taché, "Peripheral Ghrelin Selectively Increases Fos Expression in Neuropeptide Y–Synthesizing Neurons in Mouse Hypothalamic Arcuate Nucleus," *Neuroscience Letters* 325, no. 1 (2002): 47–51.

15. Z. B. Andrews, Z. W. Liu, N. Walllingford, D. M. Erion, E. Borok, J. M. Friedman, M. H. Tschöp, M. Shanabrough, G. Cline, G. I. Shulman, A. Coppola, X. B. Gao, T. L. Horvath, and S. Diano, "UCP2 Mediates Ghrelin's Action on NPY/AgRP Neurons by Lowering Free Radicals," *Nature* 454, no. 7206 (2008): 846–851.

16. C. Davis and S. Kaptein, "Anorexia Nervosa with Excessive Exercise: A Phenotype with Close Links to Obsessive-Compulsive Disorder," *Psychiatry Research* 142, nos. 2–3 (2006): 209–217.

12. VITAMIN D AND THE BRAIN: MORE GOOD NEWS

1. M. F. Holick and T. C. Chen, "Vitamin D Deficiency: A Worldwide Problem with Health Consequences," *American Journal of Clinical Nutrition* 87, no. 4 (2008): 1080S–1086S.

2. R. M. Lucas, A. L. Ponsonby, J. A. Pasco, and R. Morley, "Future Health Implications of Prenatal and Early-Life Vitamin D Status," *Nutrition Reviews* 66, no. 12 (2008): 710–720.

3. H. L. Chen and D. M. Panchision, "Concise Review: Bone Morphogenetic Protein Pleiotropism in Neural Stem Cells and Their Derivatives: Alternative Pathways, Convergent Signals," *Stem Cells* 25, no. 1 (2007): 63–68.

4. T. J. Wang, M. J. Pencina, S. L. Booth, P. F. Jacques, E. Ingelsson, K. Lanier, E. J. Benjamin, R. B. D'Agostino, M. Wolf, and R. S. Vasan, "Vitamin D Deficiency and Risk of Cardiovascular Disease," *Circulation* 117, no. 4 (2008): 503–511.

5. S. Pilz, H. Dobnig, J. E. Fischer, B. Wellnitz, U. Seelhorst, B. O. Boehm, and W. März, "Low Vitamin D Levels Predict Stroke in Patients Referred to Coronary Angiography," *Stroke* 39, no. 9 (2008): 2611–2613.

6. M. L. Melamed, E. D. Michos, W. Post, and B. Astor, "25-hydroxyvitamin D Levels and the Risk of Mortality in the General Population," *Archives of Internal Medicine* 168, no. 15 (2008): 1629–1637.

7. E. P. Cherniack, H. Florez, B. A. Roos, B. R. Troen, and S. Levis, "Hypovitaminosis D in the Elderly: From Bone to Brain," *The Journal of Nutrition, Health and Aging* 12, no. 6 (2008): 366–373.

8. M. F. Holick, "Vitamin D—New Horizons for the 21st Century," *The American Journal of Clinical Nutrition* 60, no. 4 (1994): 619–630.

9. E. P. Cherniack, B. R. Troen, H. J. Florez, B. A. Roos, and S. Levis, "Some New Food for Thought: The Role of Vitamin D in the Mental Health of Older Adults," *Current Psychiatry Reports* 11, no. 1 (2009): 12–19.

10. S. V. Ramagopalan, N. J. Maugeri, L. Handunnetthi, M. R. Lincoln, S. M. Orton, D. A. Dyment, G. C. DeLuca, B. M. Herrera, M. J. Chao, A. D. Sadovnick, G. C. Ebers, and J. C. Knight, "Expression of the Multiple Sclerosis–Associated MHC Class II Allele HLA-DRB1*1501 Is Regulated by Vitamin D," *PLoS Genetics* 5, no. 2 (2009): e1000369.

11. S. J. Kiraly, M. A. Kiraly, R. D. Hawe, and N. Makhani, "Vitamin D as a Neuroactive Substance: Review," *Scientific World Journal* 6 (2006): 125–139.

12. J. J. Cannell and B. W. Hollis, "Use of Vitamin D in Clinical Practice," *Alternative Medicine Review* 13, no. 1 (2008): 6–20.

13. L. A. Houghton and R. Vieth, "The Case Against Ergocalciferol (Vitamin D2) as a Vitamin Supplement," *American Journal of Clinical Nutrition* 84 (2006): 694–697.

13. RELIGON AND THE BRAIN: A DEBATE

1. D. Kapogiannis, A. K. Barbey, M. Su, G. Zamboni, F. Krueger, and J. Grafman, "Cognitive and Neural Foundations of Religious Belief," *Proceedings of the National Academy of Sciences USA* 106, no. 12 (2009): 4876–4881.

2. R. Varley and M. Siegal, "Evidence for Cognition without Grammar from Causal Reasoning and 'Theory of Mind' in an Agrammatic Aphasic Patient," *Current Biology* 10, no. 12 (2000): 723–726.

3. W. Schultz, "Multiple Dopamine Functions at Different Time Courses," *Annual Review of Neuroscience* 30 (2007): 259–288.

4. W. Schultz, "Behavioral Dopamine Signals," *Trends in Neurosciences* 30, no. 5 (2007): 203–210.

5. J. Yacubian, J. Glascher, K. Schroeder, T. Sommer, D. F. Braus, and C. Buchel, "Dissociable Systems for Gain- and Loss-Related Value Predictions and Errors of Prediction in the Human Brain," *Journal of Neuroscience* 26, no. 37 (2006): 9530–9537.

6. J. B. Rosen and M. P. Donley, "Animal Studies of Amygdala Function in Fear and Uncertainty: Relevance to Human Research," *Biological Psychology* 73, no. 1 (2006): 49–60.

7. A. Newberg and B. Lee, "The Neuroscientific Study of Religious and Spiritual Phenomena: Or Why God Doesn't Use Biostatistics," *Zygon* 40, (2005): 469-489.

8. R. R. Griffiths, W. A. Richards, U. McCann, and R. Jesse, "Psilocybin Can Occasion Mystical-type Experiences Having Substantial and Sustained Personal Meaning and Spiritual Significance," *Psychopharmacology* 187, (2006): 268–283.

9. H. G. Koenig, eds., *Handbook of Religion and Mental Health* (San Diego: Academic Press, 1998).

15. SYNESTHESIA:
ANOTHER WORLD OF PERCEPTION

1. J. E. Asher, J. A. Lamb, D. Brocklebank, J. B. Cazier, E. Maestrini, L. Addis, M. Sen, S. Baron-Cohen, and A. P. Monaco, "A Whole-genome Scan and Fine-mapping Linkage Study of Auditory-visual Synesthesia Reveals Evidence of Linkage to Chromosomes 2q24, 5q33, 6p12, and 12p12," *American Journal of Human Genetics* 84, no. 2 (2009): 279–285.

2. D. Maurer, "Neonatal Synesthesia: Implications for the Processing of Speech and Faces," in *Synaesthesia: Classic and Contemporary Readings*, ed. S. Baron-Cohen and J. Harrison, 224–242 (Malden, Massachusetts: Blackwell Publishers Inc., 1997).

3. H. Neville, "Developmental Specificity in Neurocognitive Development in Humans," in *The Cognitive Neurosciences*, ed. M. Gazzaniga, 219–234 (Cambridge, Massachusetts: The MIT Press, 1995).

4. D. Smilek, M. J. Dixon, and P. M. Merikle, "Synaesthesia: Discordant Male Monozygotic Twins," *Neurocase* 11 (2005): 363–370.

5. D. Smilek, B. A. Moffatt, J. Pasternak, B. N. White, M. J. Dixon, and P. M. Merikle, "Synaesthesia: A Case Study of Discordant Monozygotic Twins," *Neurocase* 8 (2002): 338–342.

6. J. Simner, C. Mulvenna, N. Sagiv, E. Tsakanikos, S. A. Witherby, C. Fraser, K. Scott, and J. Ward, "Synaesthesia: The Prevalence of Atypical Cross-modal Experiences," *Perception* 35, no. 8 (2006): 1024–1033.

7. S. Baron-Cohen, M. A. Wyke, and C. Binnie, "Hearing Words and Seeing Colours: An Experimental Investigation of a Case of Synaesthesia," *Perception* 16 (1987): 761–767.

16. WEIGHING IN ON "CONDITIONED HYPEREATING"

1. M. B. Schwartz and K. D. Brownell, "Action Necessary to Prevent Childhood Obesity: Creating the Climate for Change," *Journal of Law and Medical Ethics* 35, no. 1 (2007): 78–89.

2. T. G. Randolph, "The Descriptive Features of Food Addiction: Addictive Eating and Drinking," *Quarterly Journal of Studies of Alcohol* 17, no. 2 (1956): 198–224.

3. M. S. Gold, N. A. Graham, J. A. Cocores, and S. J. Nixon, "Food Addiction?" *Journal of Addiction Medicine* 3, no. 1 (2009): 42–45.

4. G. J. Wang, N. D. Volkow, P. K. Thanos, and J. S. Fowler, "Imaging of Brain Dopamine Pathways: Implications for Understanding Obesity," *Journal of Addiction Medicine* 3, no. 1 (2009): 8–18.

5. B. G. Hoebel, N. M. Avena, M. E. Bocarsly, and P. Rada, "Natural Addiction: A Behavioral and Circuit Model Based on Sugar Addiction in Rats," *Journal of Addiction Medicine* 3, no. 1 (2009): 33–41.

Index

A

ablative surgery, 104
abulia, 178
action observation network (AON)
    and brain injuries, 32–33
    and learning, 29–30
    and motivation, 31
    overview, 25–27
    and physical mastery, 30–31
    and rapid training, 27–28
    and social intelligence, 32
activity-dependent plasticity, 13–14
AD (Alzheimer's disease), 125
adaptive behaviors of teenagers, 63–64
adenosine triphosphate (ATP), 138
ADHD (attention-deficit/hyperactivity
    disorder), xv, 6, 10, 120
adipose tissue, 139
adolescence, 64. *See also* teen brains
adult end state and multiple intelligences
    development, 42
adults
    end state and multiple intelligences
        development, 42
    older adults' susceptibility to fraud,
        124–25, 127–28, 129–31
    and vitamin D requirements, 146–48
advertisements, deceptive and misleading,
    127–28
age of consent and maturity, 69–70
Aglioti, Salvatore, 27
agouti-related protein (AgRP), 135–36,
    137, 138, 139
alerting network of the brain, 17
Alzheimer's disease (AD), 125
Amen, Daniel, 120

American Psychiatric Association (APA),
    120. See also *Diagnostic and
    Statistical Manual of Mental
    Disorders*
amygdala, 74–75, 117
anatomic studies and connectivity, 66
anorexia nervosa, 140–42
anterior cingulate gyrus (executive
    attention network), 18
anterior portion of frontal lobes, 126
AON. *See* action observation network
APA (American Psychiatric Association),
    120. See also *Diagnostic and
    Statistical Manual of Mental
    Disorders*
appetite-stimulating and -inhibiting
    neurons, 135–36, 137, 138,
    139–40
arcuate nucleus (ARC), 135, 137
Arts and Human Development, The
    (Gardner), 47
arts training
    and activity-dependent plasticity, 13–14
    applying skills learning in, 9
    and attention networks, 13, 15, 16–19
    and cognitive function, 21–22
    data, 14–16
    individualized benefits from, 19–21
    See also dance training; music training
Asher, Julian E., 181–87
ATP (adenosine triphosphate), 138
attention-deficit/hyperactivity disorder
    (ADHD), xv, 6, 10, 120
attention networks, 13, 15, 16–19, 20,
    21–22
attention-training exercises, 17, 20
Australia, 42

Other Dana Press Books

www.dana.org/news/danapressbooks

Books for General Readers

Brain and Mind

TREATING THE BRAIN: What the Best Doctors Know

Walter G. Bradley, DM, FRCP

Using patient case histories, *Treating the Brain* explains the causes, diagnosis, prognosis, and treatment of a wide range of frequently diagnosed disorders, including Alzheimer's, migraines, stroke, epilepsy, and Parkinson's.

9 illustrations.

Cloth • 336 pp • ISBN-13: 978-1-932594-46-1 • $25.00

DEEP BRAIN STIMULATION:
A New Treatment Shows Promise in the Most Difficult Cases

Jamie Talan

An award-winning science writer has penned the first general-audience book to explore the benefits and risks of this cutting-edge technology, which is producing promising results for a wide range of brain disorders.

Cloth• 200 pp • ISBN-13: 978-1-932594-37-9 • $25.00

TRY TO REMEMBER: Psychiatry's Clash Over Meaning, Memory, and Mind

Paul R. McHugh, M.D.

Prominent psychiatrist and author Paul McHugh chronicles his battle to put right what has gone wrong in psychiatry. McHugh takes on such controversial subjects as "recovered memories," multiple personalities, and the overdiagnosis of PTSD.

Cloth • 300 pp • ISBN-13: 978-1-932594-39-3 • $25.00

CEREBRUM 2009: Emerging Ideas in Brain Science

Foreword by Thomas R. Insel, M.D.

Why does mental fuzziness follow heart surgery? Can brain scans predict how you'll vote? How life-threatening is hidden brain injury? Leading scientists and writers tackle these and other challenging issues.

Paper • 188 pp • ISBN-13: 978-1-932594-44-7 • $14.95

CEREBRUM 2008: Emerging Ideas in Brain Science

Foreword by Carl Zimmer

Is free will an illusion? Why must we remember the past to envision the future? How can architecture help Alzheimer's patients? This edition presents these and 10 other topics.

Paper • 225 pp • ISBN-13: 978-1-932594-33-1 • $14.95

CEREBRUM 2007: Emerging Ideas in Brain Science

Foreword by Bruce S. McEwen, Ph.D.

How dangerous is adult sleepwalking? Is happiness hard-wired? Could an elephant be the next Picasso? Prominent neuroscientists and other thinkers explore a year's worth of topics.

Paper • 243 pp • ISBN-13: 978-1-932594-24-9 • $14.95

Visit Cerebrum online at www.dana.org/news/cerebrum.

YOUR BRAIN ON CUBS: Inside the Heads of Players and Fans

Dan Gordon, Editor

Our brains light up with the rush that accompanies a come-from-behind win—and the crush of a disappointing loss. Brain research also offers new insight into how players become experts. Neuroscientists and science writers explore these topics and more in this intriguing look at talent and triumph on the field and our devotion in the stands.

6 illustrations.

Cloth • 150 pp • ISBN-13: 978-1-932594-28-7 • $19.95

THE NEUROSCIENCE OF FAIR PLAY:
Why We (Usually) Follow the Golden Rule

Donald W. Pfaff, Ph.D.

A distinguished neuroscientist presents a rock-solid hypothesis of why humans across time and geography have such similar notions of good and bad, right and wrong.

10 illustrations.

Cloth • 234 pp • ISBN-13: 978-1-932594-27-0 • $20.95

BEST OF THE BRAIN FROM SCIENTIFIC AMERICAN:
Mind, Matter, and Tomorrow's Brain

Floyd E. Bloom, M.D., Editor

Top neuroscientist Floyd E. Bloom has selected the most fascinating brain-related articles from *Scientific American* and *Scientific American Mind* since 1999 in this collection.

30 illustrations.

Cloth • 300 pp • ISBN-13: 978-1-932594-22-5 • $25.00

MIND WARS: Brain Research and National Defense

Jonathan D. Moreno, Ph.D.

A leading ethicist examines national security agencies' work on defense applications of brain science, and the ethical issues to consider.

Cloth • 210 pp • ISBN-10: 1-932594-16-7 • $23.95

THE DANA GUIDE TO BRAIN HEALTH:
A Practical Family Reference from Medical Experts (with CD-ROM)

Floyd E. Bloom, M.D., M. Flint Beal, M.D., and David J. Kupfer, M.D., Editors

Foreword by William Safire

A complete, authoritative, family-friendly guide to the brain's development, health, and disorders.

16 full-color pages and more than 200 black-and-white drawings.

Paper (with CD-ROM) • 733 pp • ISBN-10: 1-932594-10-8 • $25.00

THE CREATING BRAIN: The Neuroscience of Genius

Nancy C. Andreasen, M.D., Ph.D.

A noted psychiatrist and best-selling author explores how the brain achieves creative breakthroughs, including questions such as how creative people are different and the difference between genius and intelligence.

33 illustrations/photos.

Cloth • 197 pp • ISBN-10: 1-932594-07-8 • $23.95

THE ETHICAL BRAIN

Michael S. Gazzaniga, Ph.D.

Explores how the lessons of neuroscience help resolve today's ethical dilemmas, ranging from when life begins to free will and criminal responsibility.

Cloth • 201 pp • ISBN-10: 1-932594-01-9 • $25.00

A GOOD START IN LIFE:
Understanding Your Child's Brain and Behavior from Birth to Age 6

Norbert Herschkowitz, M.D., and Elinore Chapman Herschkowitz

The authors show how brain development shapes a child's personality and behavior, discussing appropriate rule-setting, the child's moral sense, temperament, language, playing, aggression, impulse control, and empathy.

13 illustrations.

Cloth • 283 pp • ISBN-10: 0-309-07639-0 • $22.95
Paper (Updated with new material) • 312 pp • ISBN-10: 0-9723830-5-0 • $13.95

BACK FROM THE BRINK:
How Crises Spur Doctors to New Discoveries about the Brain

Edward J. Sylvester

In two academic medical centers, Columbia's New York Presbyterian and Johns Hopkins Medical Institutions, a new breed of doctor, the neurointensivist, saves patients with life-threatening brain injuries.

16 illustrations/photos.

Cloth • 296 pp • ISBN-10: 0-9723830-4-2 • $25.00

THE BARD ON THE BRAIN:
Understanding the Mind Through the Art of Shakespeare and the Science of Brain Imaging

Paul M. Matthews, M.D., and Jeffrey McQuain, Ph.D. • *Foreword by Diane Ackerman*

Explores the beauty and mystery of the human mind and the workings of the brain, following the path the Bard pointed out in 35 of the most famous speeches from his plays.

100 illustrations.

Cloth • 248 pp • ISBN-10: 0-9723830-2-6 • $35.00

STRIKING BACK AT STROKE: A Doctor-Patient Journal

Cleo Hutton and Louis R. Caplan, M.D.

A personal account, with medical guidance from a leading neurologist, for anyone enduring the changes that a stroke can bring to a life, a family, and a sense of self.

15 illustrations.

Cloth • 240 pp • ISBN-10: 0-9723830-1-8 • $27.00

UNDERSTANDING DEPRESSION:
What We Know and What You Can Do About It

J. Raymond DePaulo, Jr., M.D., and Leslie Alan Horvitz

Foreword by Kay Redfield Jamison, Ph.D.

What depression is, who gets it and why, what happens in the brain, troubles that come with the illness, and the treatments that work.

Cloth • 304 pp • ISBN-10: 0-471-39552-8 • $24.95
Paper • 296 pp • ISBN-10: 0-471-43030-7 • $14.95

KEEP YOUR BRAIN YOUNG:
The Complete Guide to Physical and Emotional Health and Longevity

Guy M. McKhann, M.D., and Marilyn Albert, Ph.D.

Every aspect of aging and the brain: changes in memory, nutrition, mood, sleep, and sex, as well as the later problems in alcohol use, vision, hearing, movement, and balance.

Cloth • 304 pp • ISBN-10: 0-471-40792-5 • $24.95
Paper • 304 pp • ISBN-10: 0-471-43028-5 • $15.95

THE END OF STRESS AS WE KNOW IT

Bruce S. McEwen, Ph.D., with Elizabeth Norton Lasley • Foreword by Robert Sapolsky

How brain and body work under stress and how it is possible to avoid its debilitating effects.

Cloth • 239 pp • ISBN-10: 0-309-07640-4 • $27.95
Paper • 262 pp • ISBN-10: 0-309-09121-7 • $19.95

IN SEARCH OF THE LOST CORD:
Solving the Mystery of Spinal Cord Regeneration

Luba Vikhanski

The story of the scientists and science involved in the international scientific race to find ways to repair the damaged spinal cord and restore movement.

21 photos; 12 illustrations.

Cloth • 269 pp • ISBN-10: 0-309-07437-1 • $27.95

THE SECRET LIFE OF THE BRAIN

Richard Restak, M.D. • Foreword by David Grubin

Companion book to the PBS series of the same name, exploring recent discoveries about the brain from infancy through old age.

Cloth • 201 pp • ISBN-10: 0-309-07435-5 • $35.00

THE LONGEVITY STRATEGY:
How to Live to 100 Using the Brain-Body Connection

David Mahoney and Richard Restak, M.D. • Foreword by William Safire

Advice on the brain and aging well.

Cloth • 250 pp • ISBN-10: 0-471-24867-3 • $22.95
Paper • 272 pp • ISBN-10: 0-471-32794-8 • $14.95

STATES OF MIND: New Discoveries About How Our Brains Make Us Who We Are

Roberta Conlan, Editor

Adapted from the Dana/Smithsonian Associates lecture series by eight of the country's top brain scientists, including the 2000 Nobel laureate in medicine, Eric Kandel.

Cloth • 214 pp • ISBN-10: 0-471-29963-4 • $24.95

Paper • 224 pp • ISBN-10: 0-471-39973-6 • $18.95

The Dana Foundation Series on Neuroethics

DEFINING RIGHT AND WRONG IN BRAIN SCIENCE:
Essential Readings in Neuroethics

Walter Glannon, Ph.D., Editor

The fifth volume in The Dana Foundation Series on Neuroethics, this collection marks the five-year anniversary of the first meeting in the field of neuroethics, providing readers with the seminal writings on the past, present, and future ethical issues facing neuroscience and society.

Cloth • 350 pp • ISBN-10: 978-1-932594-25-6 • $15.95

HARD SCIENCE, HARD CHOICES:
Facts, Ethics, and Policies Guiding Brain Science Today

Sandra J. Ackerman, Editor

Top scholars and scientists discuss new and complex medical and social ethics brought about by advances in neuroscience. Based on an invitational meeting co-sponsored by the Library of Congress, the National Institutes of Health, the Columbia University Center for Bioethics, and the Dana Foundation.

Paper • 152 pp • ISBN-10: 1-932594-02-7 • $12.95

NEUROSCIENCE AND THE LAW: Brain, Mind, and the Scales of Justice

Brent Garland, Editor. With commissioned papers by Michael S. Gazzaniga, Ph.D., and Megan S. Steven; Laurence R. Tancredi, M.D., J.D.; Henry T. Greely, J.D.; and Stephen J. Morse, J.D., Ph.D.

How discoveries in neuroscience influence criminal and civil justice, based on an invitational meeting of 26 top neuroscientists, legal scholars, attorneys, and state and federal judges convened by the Dana Foundation and the American Association for the Advancement of Science.

Paper • 226 pp • ISBN-10: 1-032594-04-3 • $8.95

BEYOND THERAPY: Biotechnology and the Pursuit of Happiness
A Report of the President's Council on Bioethics

Special Foreword by Leon R. Kass, M.D., Chairman

Introduction by William Safire

Can biotechnology satisfy human desires for better children, superior performance, ageless bodies, and happy souls? This report says these possibilities present us with profound ethical challenges and choices. Includes dissenting commentary by scientist members of the Council.

Paper • 376 pp • ISBN-10: 1-932594-05-1 • $10.95

NEUROETHICS: Mapping the Field. Conference Proceedings

Steven J. Marcus, Editor

Proceedings of the landmark 2002 conference organized by Stanford University and the University of California, San Francisco, and sponsored by the Dana Foundation, at which more than 150 neuroscientists, bioethicists, psychiatrists and psychologists, philosophers, and professors of law and public policy debated the ethical implications of neuroscience research findings.

50 illustrations.

Paper • 367 pp • ISBN-10: 0-9723830-0-X • $10.95

Immunology

RESISTANCE: The Human Struggle Against Infection

Norbert Gualde, M.D., translated by Steven Rendall

Traces the histories of epidemics and the emergence or re-emergence of diseases, illustrating how new global strategies and research of the body's own weapons of immunity can work together to fight tomorrow's inevitable infectious outbreaks.

Cloth • 219 pp • ISBN-10: 1-932594-00-0 • $25.00

FATAL SEQUENCE: The Killer Within

Kevin J. Tracey, M.D.

An easily understood account of the spiral of sepsis, a sometimes fatal crisis that most often affects patients fighting off nonfatal illnesses or injury. Tracey puts the scientific and medical story of sepsis in the context of his battle to save a burned baby, a sensitive telling of cutting-edge science.

Cloth • 231 pp • ISBN-10: 1-932594-06-X • $23.95
Paper • 231 pp • ISBN-10: 1-932594-09-4 • $12.95

Arts Education

A WELL-TEMPERED MIND: Using Music to Help Children Listen and Learn

Peter Perret and Janet Fox • Foreword by Maya Angelou

Five musicians enter elementary school classrooms, helping children learn about music and contributing to both higher enthusiasm and improved academic performance. This charming story gives us a taste of things to come in one of the newest areas of brain research: the effect of music on the brain.

12 illustrations.

Cloth • 225 pp • ISBN-10: 1-932594-03-5 • $22.95
Paper • 225 pp • ISBN-10: 1-932594-08-6 • $12.00

Dana Press also publishes periodicals dealing with brain science, arts education, and immunology. For more information, visit www.dana.org.